Water, Stones, & Fossil Bones

Copyright © 1991 by the National Science Teachers Association, 1742 Connecticut Avenue, NW, Washington, DC 20009. Permission is granted in advance for reproduction for the purpose of classroom or workshop instruction.

This book has been edited and produced by the staff of NSTA Special Publications. Shirley Watt Ireton, Managing Editor; Christine M. Pearce, Assistant Editor; Andrew Saindon, Assistant Editor; Gregg Sekscienski, Editorial Assistant. The book design is by Ellen Baker of Auras Design and the cover was illustrated by Tim Knepp.

Library of Congress Catalog Card Number 91-60991

Stock Number PB–89

ISBN Number 0-87355-101-x

Printed in the United States of America

Water, Stones, & Fossil Bones

Earth science activities for elementary and middle level grades

Edited by Karen K. Lind
University of Louisville
Louisville, Kentucky

A joint publication of

Council for Elementary Science International

and

National Science Teachers Association

CESI Sourcebook VI

CONTENTS

Acknowledgements vii
Foreword viii
Preface ix
Introduction xi

Space

1 Scale Model Solar System 2
Michael Kotar

2 Moon Loops 4
Michael Kotar

3 Earth Alignment in Space 6
Michael Kotar

Land

4 Igneous Fudge 10
Karen K. Lind

5 Exploring Cookies 12
Karen K. Lind

6 Edible Conglomerates 14
Christy D. McGee

7 Play-Doh Stratigraphy 16
Lee A. Krohn

8 Collect and Compare 18
Rose West

9 Smaller than Your Pinky, Larger than Your Fist 20
Betty B. Graham

10 Rock and Roll 23
Paul Caulfield and Rose West

11 Weathering or Not? 25
Shirley G. Key

12 Dirt Cake 27
Karen K. Lind

13 Shake and Sieve! 29
Sue Dale Tunnicliffe

Land (continued)

14 Soil Profiles 32
David R. Stronck

15 Big and Little Sand 34
Lloyd H. Barrow

16 Mini Landfills 35
Mildred Moseman

17 A Model of the Earth's Crust 37
Gerald Wm. Foster

18 Model of a Seismograph 39
Muhammad Hanif

19 Songs of Earth Science 42
Rose West

Water

20 How Low Does It Go? 46
Susan M. Johnson

21 Soil Percolating 48
Michael J. Demchik

22 River Boxes 50
David R. Stronck

23 Clean Water: Is It Drinkable? 52
Carol VanDeWalle

24 Fred the Fish 54
Patricia Chilton-Stringham and Jan Wolanin

25 Soil Leaching 58
Gerald Wm. Foster

26 When It Rains, It Pours 61
M. Elizabeth Partridge

27 Grinding and Scraping 63
Karen K. Lind

28 Drip Sculpture 65
Edward P. Ortleb

Air

29 Hot Days, Cold Days 68
Maureen J. Awbrey 68

30 Capturing Heat from the Sun 70
Vincent G. Sindt

31 Cooking with the Sun 74
Ruth M. Ruud

32 Catch a Raindrop 76
Gerald Wm. Foster

33 Fakey Fog? 78
Gerald Wm. Foster

34 A Simple Hygrometer Mobile 80
Denise Tassi-Kane

35 Moving Air—The Whoosh Box 82
Judy Pessolano

36 When Is It Dew? 84
Shirley G. Key

37 Dew or Frost? 86
Merrick Owen

38 Something in the Air 87
M. Elizabeth Partridge

39 Acid Precipitation 89
Michael J. Demchik

The Earth's Past

40 Ancient Earth 92
Amy Lowen and Lisa Weaver

41 The Half-life and Times of Geologic Materials 94
Kevin D. Finson

42 Timelining How Old Is Old? 96
Charles R. Ault, Jr.

43 Magnetic Tracks 100
Betty B. Graham

44 What Is a Fossil? 102
Mildred Moseman

45 Future Fossils 104
Gerald Wm. Foster

46 Fossil Beds 106
David R. Stronck

47 Double-time, Clean Up!! 108
Larry Flick

48 Hands-on Mapping 111
John B. Beaver and Michael G. Jacobson

49 The Great Flood 113
Carole J. Reesink

50 How Much Is a Million? 117
Larry Flick

51 Litter Alert 120
Robert N. Ronau

Council for Elementary Science International 124

Acknowledgements

Many thanks to the CESI members who found time to tell about their favorite Earth science activities. Teachers sharing with other teachers is one of the strengths of CESI and of elementary school teaching. This sourcebook clearly demonstrates that an organization of teachers, presenting publications by and for teachers, is a valuable resource.

I wish to express my appreciation to my family—Eugene Lind, Pamela, Paul, and Marian Kalbfleisch of Filer, Idaho—for their encouragement. Thanks also to the University of Louisville for supporting the sourcebook; typists Linda Moore, Robyn Roberts, and Kathy Flaker for their work; doctoral student Christy McGee for her concern and assistance. A note of praise goes to Phyllis Marcuccio for advice and encouragement and to reviewers Robert Ronau, Katherine Becker, Jeff Callister, M. Frank Watt Ireton, Sheila Marshall, Jim Sproull, and Steve Vandas for their attention to accuracy. Special thanks to NSTA staff Shirley Watt Ireton, Christine M. Pearce, and Pat Tschirhart-Spangler for helping us craft our strategies into a book.

—Karen K. Lind

Illustration Credits

p. 10   Karen K. Lind

p. 12   Reprinted, by permission, from *Math and Science for Young Children* (p. 490) by R. Charlesworth and K. Lind, 1990, Albany, NY: Delmar Publishers. © 1990 by Delmar Publishers.

p. 27   Karen K. Lind

p. 56   Benjamin Matthew West drew this in 1987–1988 as a 5th grade student of contributing author Jan Wolanin at St. Francis School in Goshen, KY.

p. 69   Andrew Saindon

p. 80   Reprinted, by permission, from *Make a simple hygrometer* by Franklin Institute, 1986, Philadelphia, PA: Author

p. 115 Carole Reesink

All other illustrations were created by Cynthia Cliff.

FOREWORD

GERALD H. KROCKOVER
PROFESSOR OF EDUCATION
AND GEOSCIENCES
PURDUE UNIVERSITY
WEST LAFAYETTE, INDIANA

Why Teach Earth Science?

Why teach Earth science in elementary school? Because children see the subject all around them. Earth science encompasses all sciences, including such familiar topics as weather, climate, rocks and minerals, rivers and streams, astronomy. Its interdisciplinary nature builds upon the life, physical, and mathematical sciences; tying the curriculum together with material drawn from language arts, social studies, and mathematics.

The Earth sciences lend themselves to the theme approach to learning. Themes such as *change* or the *environment* build upon fundamental Earth science concepts. Teach Earth science with the theme of change and you provide students with the opportunity to directly observe physical and chemical changes. Change can also be taught using learning centers, each focusing upon one aspect of change. A study of fossils can illustrate grand change—evolution. Weather provides its own classic example of change, both its causes and its effects on rocks and minerals.

Teach Earth science with an environmental theme and you provide opportunities for students to investigate interdisciplinary topics such as recycling, pollution, landfills, extinction, global warming, space, mining, outdoor areas, national parks, waterways, and oil spills.

Earth science is exciting for children. Consider the words earthquake, flood, hurricane, landslide, volcano, tornado, erosion, pollution, dinosaur, universe, space station—all Earth science words with adventure potential. Once children's enthusiasm is aroused they are eager to learn. Your next step is to provide them with avenues for exploration; avenues that you can find in *Water, Stones, & Fossil Bones*.

The Learning Cycle and Earth Science

During the 1960s, the authors of an elementary science program known as Science Curriculum Improvement Study (SCIS) designed a teaching approach called the learning cycle. In evaluating the SCIS program and in studies at other grade levels, researchers have repeatedly shown the learning cycle to be an effective teaching strategy.

CHARLES BARMAN
ASSOCIATE PROFESSOR OF
SCIENCE EDUCATION
INDIANA UNIVERSITY/PURDUE
UNIVERSITY AT INDIANAPOLIS

The learning cycle has three distinct phases: exploration, concept introduction, and concept application.

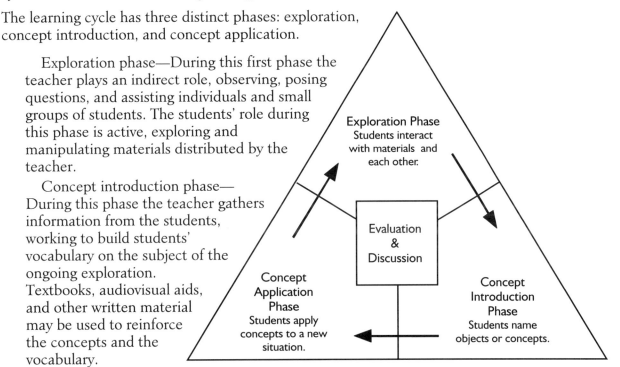

> Exploration phase—During this first phase the teacher plays an indirect role, observing, posing questions, and assisting individuals and small groups of students. The students' role during this phase is active, exploring and manipulating materials distributed by the teacher.
>
> Concept introduction phase—During this phase the teacher gathers information from the students, working to build students' vocabulary on the subject of the ongoing exploration. Textbooks, audiovisual aids, and other written material may be used to reinforce the concepts and the vocabulary.
>
> Concept application phase—Now the teacher poses a new situation or problem which students can solve by extrapolating the previous exploration and its concepts. The concept application phase resembles the exploration phase in the physical activity of the students.

The three phases of the learning cycle follow a pattern that helps students develop new concepts. The exploration phase provides students with concrete experiences on which to build mental images of the new ideas or terms presented in the concept introduction phase. Then the concept application phase provides students with an opportunity to use these new ideas or terms in different situations, increasing their chances of internalizing them.

NSTA and CESI designed *Water, Stones, & Fossil Bones* to provide you with a variety of activities that can be presented using the learning cycle. Use the Challenge section to introduce the exploration phase; the Procedure will provide a structure for the students' exploration. Focus sections provide a first source for the terms and concepts you wish students to understand, though you will want to augment with trade books, videos, textbooks, and other material. Each activity has a Further Challenges section for use during the concept application phase. If you and your students find yourselves intrigued by another aspect of the exploration that could be used to cement concepts during this last phase, proceed! Satisfying curiosity is the cornerstone of science and learning in general.

INTRODUCTION

Using These Earth Science Activities

KAREN K. LIND
ASSISTANT PROFESSOR OF EARLY
AND MIDDLE CHILDHOOD
EDUCATION
UNIVERSITY OF LOUISVILLE

As a teacher you play a primary role in establishing students' attitudes toward science. By creating a positive environment for learning and providing activities your students get physically and mentally involved in, you foster the natural curiosity of students. For example, students' curiosity about rain, snow, and clouds is the beginning of their learning specific skills such as reading weather charts, as well as learning general concepts about weather. Making science fun is essential to stimulate students to learn and apply what they learn to their lives.

But how do you make science fun? That is what *Water, Stones, & Fossil Bones* can show you! In addition to providing teacher-tested, hands-on activities that work, this book stresses the teacher's role in selecting appropriate activities and asking questions that guide students to higher-order thinking. Questions to ask during the activity are suggested in margin notes, as are such teaching strategies.

The five section introductions present additional teaching strategies, tips, and information that will make your teaching of science more effective. The topics of these section introductions offer techniques for teaching process skills, applying cognitive development knowledge, managing exploration, developing teacher questioning methods, and showing the relationship between language development and thinking skills. Each introduction is related to the activities in that section and gives you examples for each teaching strategy.

The activities in this book include the following information:

Focus: a short description of the concept or skills developed by the activity and some background information on the topic of interest.

Challenge: a question or problem related to the activity that could be used to initiate discussion.

Materials and Equipment: a list of all items necessary to conduct the activity. Materials are limited to those easily obtained from local sources.

Procedure: suggestions for conducting the activity with students. When appropriate, suggestions are made for introducing and concluding the activity, grouping students, and handling equipment.

Further Challenges: possible variations or extensions of the activity. Add your own!

References: articles or books that relate to the activity or provide additional background information.

Earth Science—Process Skills

Process skills are thinking skills that we use to process information, think through problems, and formulate conclusions. They are basic to thinking about and investigating the content of science. By teaching students these important skills, we enable them to learn about their world. Listed below are the basic process skills.

Observing—Using the senses to gather information. Teaching strategies that reinforce observation skills require students to watch carefully to note specific phenomena that they might ordinarily overlook.

Classifying—Grouping and sorting according to categories, such as color or use.

Measuring—Describing something in quantitative terms. Incorporate measuring into lessons by having students measure for a reason, such as "How much paper do we need to make a scale model of the solar system."

Communicating—Communicating ideas, directions, and descriptions in written form or orally so that others can understand.

Inferring—Recognizing patterns based on observation, and using these patterns to draw more meaning from a situation than can be directly observed.

Predicting—Closely related to inferring, predicting is making reasonable guesses about what will happen based on observations.

The study of Earth science provides many opportunities to develop process skills. In "Earth Alignment" students determine true north by tracking a shadow, and then infer the Earth's orientation in space. They develop spatial abilities and the skills of communicating, predicting, and inferring. This and many of the activities in this book emphasize the thinking skills that scientists use to answer questions and solve problems.

1 Scale Model Solar System

BY MICHAEL KOTAR

Focus:
Distances in space, even within the limits of our own solar system, can be incomprehensible to students. If you want to amaze your students, this activity will give you a model of the solar system in which distance and planet sizes are modeled to actual proportions. Even a portion of this model will give an excellent representation of the immensity of the solar system.

Challenge:
How much room will it take to create a scale model of the solar system? Can you make one that includes the orbit of Pluto?

Time: 30 minutes

Procedure:

1. Glue the planet models to index cards and label the cards with the planet's name. The table below lists the distances and planet sizes for this model using a scale of 1 m = 4 million km, or 1 mm = 4,000 km. With this scale, the diameter of the Sun is 29 cm and can be represented by a ball about the size of a large playground ball or simply by a circle 29 cm in diameter cut out of poster paper.

Materials and Equipment:

The whole class will need:

Meter sticks or metric measuring tape

A ball or circle 29 cm in diameter for the Sun

Index cards

Different sized small balls, lead shot, and BBs to represent the planets (see table on the next page for sizes) for example:

A ping-pong ball for Jupiter

1 mm lead shot for Mercury

2 mm lead shot for Mars

Copper BBs for Earth and Venus

Solar System to Scale: 1 m = 4 million km or 1 mm = 4,000 km

	Mercury	Venus	Earth	Mars	Jupiter	Saturn	Uranus	Neptune	Pluto
Actual									
Distance from Sun (millions of km)	58	108	150	228	778	1,429	2,875	4,504	5,900
Diameter (in km)	4,878	12,104	12,756	6,796	139,823	116,464	50,700	48,900	3,000
To Scale									
Distance from Sun (in meters)	15	27	38	57	195	357	719	1,126	1,475
Diameter (in mm)	1	3	3	2	35	29	13	12	0.8

2. Part of this solar system model can be set up in a long hallway, on a playground, or along a sidewalk. Measure the distance from the Sun, and tape the planet cards to the wall or on sticks at the scale distances from the Sun. As each planet is placed, have students look back to the ball representing the Sun. Emphasize the space encompassed by the orbit and the small size of the planets.

Q How small must you make the model planets in order to keep your model within the confines of the playground? The classroom?

The Author

Michael E. Kotar, Ed.D., is an associate professor of education at California State University in Chico.

Further Challenges:

Once the class has placed Earth or Mars at its scale distance, encourage them to think about the large amount of space encompassed by the almost circular orbits of planets in the solar system.

What are the maximum possible distances between Earth and Mars (or other planets)? What are the minimum possible distances? If the Earth's orbit is almost circular, about how much area is encompassed by its orbit?

Water, Stones, & Fossil Bones

2 Moon Loops

BY MICHAEL KOTAR

Focus:

Many people draw the Moon's orbit looping around the Earth as both travel around the Sun, but actually the annual orbit of the Moon around the Sun is similar to Earth's orbit around the Sun—almost a circle. Here is a scale model activity that clearly demonstrates this and enhances student understanding of the Earth-Moon-Sun system. As the Earth orbits the Sun, the Moon orbits the Earth. The Moon's revolution around the Earth, known as a lunar month, is approximately 28 days long, and there are about 13 lunar months in one Earth year.

Challenge:

What would the Moon's pathway around the Sun look like for one lunar month or for one year? Try to draw it from a position in space looking down on the solar system.

Time: 45 minutes

Procedure:

1. Review the Moon's phases, and have your students draw their ideas of the Moon's annual orbit around the Sun. Use a model to show the positions of the Moon, Sun, and Earth relative to each other at the full, last quarter, new, and first quarter phases of the Moon.

2. For this activity, the scale distance between the Earth and the Sun (150 million km) is 4.5 m, the length of Earth's orbit for one lunar month (72.5 million km) is 2.2 m, and the

Materials and Equipment:

The whole class will need:

Light wire or string 4.5 m long

A metal washer 2.4 cm diameter (size of a U.S. quarter)

Paper 2.2 m long and at least 20 cm wide such as tractor-feed computer paper

A pencil

A protractor

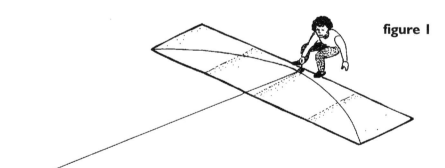

figure 1

distance between the Earth and the Moon (384 thousand km) is 12 mm, about the radius of a U.S. quarter.

3. Begin by laying the 2.2 m length of paper on the floor. Attach the washer to one end of the 4.5 m string or wire, and then lay the string or wire perpendicular to and bisecting the

paper with the washer end on the paper (see figure 1).

4. Have a student hold the other end of the string on the floor. This student is in the Sun's position in this model. Now, guide a pencil with the string and washer to make a smooth arc across the paper, from one end to the other. This represents the portion of Earth's path around the Sun for one lunar month.

5. Fold the paper in half and then in half again, dividing it into four sections representing the four weeks of the lunar month. Now, faintly draw lines equally dividing each week into seven segments.

6. On the arc, draw a dot representing the Earth in each of the 28 segments.

7. Now, mark the Moon's position for each day. The Earth/Moon distance in this scale is 12 mm. Use the 2.4-cm washer, a quarter, or similar round object to gauge the Earth/Moon distance. Start at the right side of the paper on Day 1

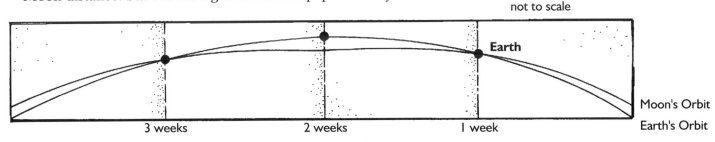

figure 2

with a full Moon. Center the washer or other round object on the Earth dot, and put a dot representing the Moon 12 mm away from the Earth. At full Moon phase, the Earth is between the Sun and the Moon.

8. Each day, the Moon's position moves about 13 degrees counterclockwise around the Earth. To make a dot for the Moon's position on each day, mark rays 90 degrees apart on the washer. Then mark the dot for Day 1 so that the Moon, Earth, and Sun are aligned. For each succeeding day estimate the 13 degree movement as 1/7 of the 90 degree angle marked on the washer. At Day 7 the Moon should be on the line representing Earth's orbit and to the left of the Earth. Day 14 will be new Moon phase, and the Moon will be between the Earth and the Sun. By Day 21, the Moon reaches first quarter phase and is on the line of Earth's orbit to the right of the Earth. Day 28 begins another full Moon phase. Take a pencil and connect the dots to see the Moon's orbit for one lunar month (see figure 2).

Q What would the Moon's orbit around the Sun for one year look like? What would the orbit of Jupiter's moons around the Sun look like?

The Author

Michael E. Kotar, Ed.D., is an associate professor of education at California State University in Chico.

Water, Stones, & Fossil Bones

3 Earth Alignment in Space

BY MICHAEL KOTAR

Focus:

The shortest daily shadow of a post or stick defines a true north line and is produced as the Sun crosses the local meridian at noon. The local meridian is the line (longitude) that runs from the North Pole to the South Pole through your location.

Once a globe is set up in its correct spatial alignment, your students will notice that your town is on the top of the globe. This positioning maintains alignment with true north and keeps the shadows of toothpicks and nearby structures parallel. No matter what your location, this will always occur, and the toothpick at your town will point straight up just as the nearby flagpole does.

Challenge:

Can you set up a globe outside so that it is aligned just as the Earth is in space relative to the Sun? Can you locate where, in the world, the Sun is rising? Where is it setting? Where is it directly overhead?

Time: Two hours to establish true north and 15 minutes to align the globe

Procedure:

1. Stick the nail through the hole in the plywood so the nail is perpendicular to the board, then place the paper over the nail, and tape it to the plywood.

2. Now, around 11 a.m., place the board in an open, sunny area so that the nail's shadow will stay on the paper for the next two hours. Every 30 minutes, mark on the paper where the tip of the nail's shadow falls. Be sure not to move the board at all until the local meridian is established. A nearby post, flagpole, or other upright structure, whose shadow can be tracked throughout the day, could also be used to locate the meridian.

3. At the end of 2 hours, around 1 p.m., connect the shadow end points, and locate the shortest distance between this line and the nail. This shortest-distance line, the local meridian, defines true north. Mark the local meridian by extending the ray from the paper to the ground with a string and two stakes or an actual chalk line on the ground.

Materials and Equipment:

The whole class will need:

An open, sunny area

A globe, preferably not attached to its base

Modeling clay

Toothpicks

2 stakes or a chalk line

A 16 penny nail

A notebook-sized piece of plywood with a hole drilled at one end for the nail

A piece of unlined 21.5 cm x 28 cm paper

4. Later in the afternoon, set up the globe to correspond to the Earth's spatial alignment toward the Sun. Place a toothpick in a small clay mound at your location on the globe. The toothpick must be perpendicular to the surface of globe. If a building or flagpole is not nearby, you may want to put a tall stake in the ground to cast a shadow with which you can align the toothpick shadow. Use a plumb bob to make sure the stake is vertical.

5. Align the north-south axis of the globe so that it is parallel to the local meridian that has been marked on the ground. Then, keeping the north-south axis aligned with the meridian, rotate the globe so the shadow cast by the toothpick is parallel to a nearby shadow of a flagpole, building, or tall stake. The globe should now be aligned as the Earth is in space. (See the figure. It represents your experiment in late afternoon.)

Further Challenges:

It is possible to discover information about Earth's alignment in space at special times by placing a toothpick at locations such as the equator or the Tropic of Cancer. After doing so, turn the globe so that the toothpick makes no shadow. You can find out, for example, how far south of the North Pole there is no night (darkness) on the first day of summer, or no daylight on the first day of winter, or how the Earth appears relative to the Sun on the vernal or autumnal equinoxes.

Q How would you describe the position of your town with respect to the rest of the Earth? Where is the Sun currently rising and setting? Where is the Sun currently directly overhead? (Hint: Use a second toothpick and mound of clay to move to the place where there is no shadow.)

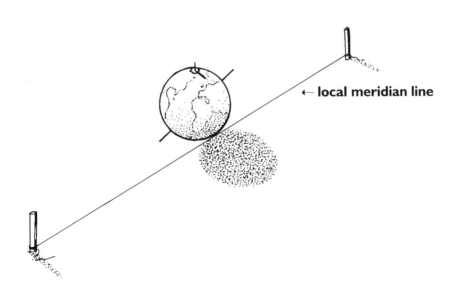
← local meridian line

Water, Stones, & Fossil Bones

The Author
Michael E. Kotar, Ed.D., is an associate professor of education at California State University in Chico.

References:

Elementary Science Study. (1971). *Teacher's guide for daytime astronomy*. New York: Webster Division, McGraw-Hill.

Schmidt, Victor (1982). *Teaching science with everyday things* (2nd ed.). New York: McGraw-Hill.

Tannenbaum, Harold E., et al. (1976). *Teacher's guide for day and night*. New York: Webster Division, McGraw-Hill.

Earth Science— At Their Own Pace

Most elementary and many middle school students are still in what Jean Piaget calls the concrete operational stage of cognitive development. While in this stage, students need to manipulate objects to grasp concepts. For example, before understanding the basic concept that soil contains a variety of particles of rock and organic materials, students need concrete, hands on experience with soil. In "Shake and Sieve," students build their own sieves and use them to sift soil. Not only do they examine the sifted soil particles, they test the effect of shaking on the size of the particles sifted and devise ways of comparing soil particles in different soil samples.

Students enter activities at their own pace and level. For example, a student who is not aware that the rate of cooling of magma affects the crystal size of the resulting rock can begin to construct this concept in the activity "Igneous Fudge." On the other hand, a student who has had experience with cooling rates can explore more advanced concepts of heating and cooling during igneous rock formation. The wide variety of Further Challenges at the end of activities in this book accommodates different levels of thinking, creativity, and development.

4 Igneous Fudge

BY KAREN K. LIND

Focus:
Igneous rocks such as pumice, obsidian, basalt, and granite harden from molten state. They have cooled at different rates. Different rates of cooling result in different size crystals. For example, large crystals, such as those found in granite, form during very slow cooling. On the other hand, fast cooling produces small crystals to form as in basalt, and very fast cooling produces glassy igneous rocks such as obsidian and pumice which do not have crystals.

Challenge:
What does the cooling rate have to do with the many different crystal sizes found in igneous rocks?

Time: 45 minutes

Procedure:
1. Have the students examine samples of obsidian, granite, and basalt and note the differences among the samples.

2. Grease the small pie tin and put enough ice into the larger pie tin so that the smaller one can easily be set into the ice (see the figure).

3. Mix the water, sugar, salt, cocoa, and vanilla.

4. Heat the mixture over a hot plate, stirring continuously, and boil it for three minutes.

Materials and Equipment:
The whole class will need:

Samples of igneous rocks such as pumice, obsidian, granite, and basalt

A hot plate

Ice

A large pie tin or wide shallow bowl

A small pie tin

Measuring spoons

Recipe for fudge:

78 ml (1/3 cup) water

237 ml (1 cup) sugar

A pinch of salt

45 ml (3 tbsp.) of cocoa

5 ml (1 tsp.) vanilla

Enough shortening or butter to grease a small pie tin

✤Safety Note:
This activity is best done as a demonstration with younger students. Even with older ones, instruct them to be extra careful around the hot plate and allow only one student at a time to heat and stir the mixture.

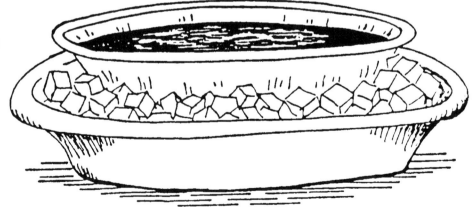

5. Turn off the hot plate, and pour half of the mixture into the greased pie pan and set it in the bowl of ice. Let the rest of the mixture cool on the burner.

6. Examine the mixtures for differences. The mixture that cooled on ice should have a smooth, glassy look and the other mixture should appear grainy.

7. Compare the obsidian, granite, and basalt to the final igneous fudge mixtures.

Further Challenges:

Prepare the fudge in a slightly different way to encompass more Earth science concepts. Repeat the activity, but layer the sugar and cocoa in a clear glass dish so students can watch the ingredients mixing together. Compare the mixing of the ingredients to the mixing of Earth's magma. Pour the fudge into three different trays, and place one in a freezer, a second tray in an oven preheated to 177°C (350°F) and turned off when the fudge is placed inside, and the third at room temperature. Let each cool completely and examine the texture of the samples.

Q What will the mixture that cools in the bowl of ice look like? What will the mixture cooling on the stove look like?

Q Why do the rocks look and feel different?

The Author

Karen K. Lind is an assistant professor in the school of education at the University of Louisville in Louisville, KY.

5 Exploring Cookies

BY KAREN K. LIND

Focus:
The Earth acts somewhat like a cook, forming rocks and minerals with pressure, heat, and motion.

Challenge:
Bake a batch of oatmeal chocolate chip cookies. Can you draw an analogy between the cookies and rocks made of different types of ingredients or minerals?

Materials and Equipment:

The whole class will need:

Mineral samples

An assortment of rocks that clearly show different crystals

Access to an oven, possibly in the school cafeteria

Two large bowls

An electric mixer

A sifter

A spatula

A wooden spoon

Measuring spoons

Teaspoons

Several large cookie sheets

Oven mitts

Cooling racks

Each student will need:

One paper towel or napkin

Time: 45 minutes

Procedure:

1. Work together as a class to make the batter and bake the cookies.

2. Display the assortment of rocks. Examine them closely, noting the variety of distinct crystals in each rock. Explain to your students that these different crystals are analogous to the different ingredients in the cookies.

3. Look at the mineral samples, and try to match these samples to crystals in the rocks.

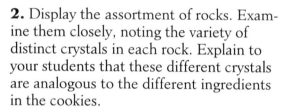

Q Which ingredients can you identify after the cookies have been baked? Can you taste the individual ingredients?

4. After the cookies have cooled, give each student two cookies, one to study and one to eat. Break open the cookies and compare them to the distinct crystals in the rocks previously examined. Explain to the students that, like their cookies, rocks have ingredients too. The Earth acts like a cook and mixes ingredients to make rocks.

5. Once the students see that some ingredients are still recognizable after baking and that some are not, ask them in what ways the ingredients have changed. The effects of heat and chemical change could be discussed at this point.

Further Challenges:

Using a rock that will show different crystals or grains, such as granite or sandstone, wrap the rock in cloth, and hit it with a hammer. Give the students hand lenses to examine the pieces of broken rock and have them examine the shape, size, and color of the fragments and describe the inside and outside appearances of the rock. Ask them why the inside looks different from the outside and how many "ingredients" they see.

References:

Lind, Karen K. (1989). Geologist for a day. *Science and Children, 26*(7), 36–37.

✢Safety Note:

Before beginning this activity, it's best to check with parents for any possible food allergies among the students.

Recipe for 5 doz. cookies:

178 ml (3/4 cup) shortening

237 ml (1 cup) brown sugar

119 ml (1/2 cup) sugar

1 egg

59 ml (1/4 cup) water

5 ml (1 tsp.) vanilla

237 ml (1 cup) flour

5 ml (1 tsp.) salt

2.5 ml (1/2 tsp.) baking soda

711 ml (3 cups) uncooked oats

One 453-g (16-oz.) package of chocolate, butterscotch, or peanut butter chips

Enough extra shortening to grease cookie sheets

With electric mixer, beat together shortening, sugars, egg, water, and vanilla until creamy. In a separate bowl, sift together flour, salt, and baking soda. Add to creamed mixture and blend well with wooden spoon. Stir in oats, then add chips and stir. Drop by rounded teaspoonfuls onto greased cookies sheets. Bake at 177°C (350°F) for 12–15 minutes. Remove from the oven and transfer to cooling racks.

The Author

Karen K. Lind is an assistant professor in the school of education at the University of Louisville in Louisville, KY.

6 Edible Conglomerates

BY CHRISTY D. MCGEE

Focus:

Conglomerates are a type of sedimentary rock made of pebbles, sand, clay, silt, and gravel cemented together by pressure and chemicals that precipitate out of water such as calcium carbonate and silica. The large size of the constituent fragments reflect the strength of the current which carried and deposited them.

Conglomerates contain pebbles of different types of rocks in a variety of shapes and sizes. Cookies can resemble real conglomerates in that they too contain fragments of various shapes and sizes (nuts, popcorn, marshmallows). In addition, conglomerates are formed under pressure and cemented together by precipitated chemicals; similarly, edible conglomerates are cemented together with pressure applied by the students' hands and a hardened sugar mixture.

Challenge:

Using popcorn, nuts, miniature marshmallows, and sugar, can you make an "edible conglomerate?" How are these edible conglomerates similar to real conglomerates? How are they different?

Time: 45 minutes

Procedure:

1. Pass around the samples of conglomerate rocks for students to hold and examine, and discuss how real conglomerate rocks form.

2. Follow the recipe for the popcorn ball cement that will hold together the different bits of the edible conglomerate. Allow the children to stir and measure the ingredients for the cement while discussing the characteristics of conglomerates.

3. Divide the class into groups of four students, and equally divide the popcorn, marshmallows, and nuts among the groups. Give each group approximately half a stick of margarine or butter to grease their hands before making their conglomerates. It helps to keep the margarine in a freezer until time for class.

4. After the sugar mixture has cooled a bit, pour equal amounts into each group's aluminum pan and have the students mix their popcorn, marshmallows, and nuts into the liquid, all of which represent the fragments and cement in a conglomerate. They should realize that although they have combined different components, they have yet to apply

Materials and Equipment:

The whole class will need:

A selection of conglomerate rocks

A large bowl

A mixing spoon

Measuring spoons and cups

Hot mitts

A hot plate

Each group of four students will need:

A disposable aluminum baking pan

Bowls

Wax paper (for each student)

Q How do conglomerates differ from other kinds of rocks? Which is older, the pebbles in the conglomerate or the conglomerate itself?

Q How does the function of the sugar cement compare to the natural cement in real conglomerates? Do the fragments in the real conglomerates show as much variation in size as the ingredients in your popcorn balls?

Q Compare what you have done so far with the formation of real conglomerate rocks.

pressure, which is needed during the formation of real conglomerates.

5. Give each student a piece of wax paper. Have the students grease their hands and press together handfuls of the mixture to complete their model conglomerates. Leave the conglomerates on the wax paper or use the wax paper to wrap the conglomerates for students to take home.

6. Compare the students' edible conglomerates to the real ones examined earlier.

Further Challenges:

Can your students suggest ways or recipes to make different types of edible rocks? What other foods can be made into edible rocks? Try your ideas out to see if they work.

✢Safety Note:

In using the hot plate, remind students to keep long sleeves and hair away from the heat source and to avoid being splattered by the hot mixture.

Recipe for 25–30 popcorn balls:

About 4.7 L (20 cups) of popped corn

474 ml (2 cups) of colored miniature marshmallows

474 ml (2 cups) of nuts

3 sticks of margarine or butter (to put on students' hands—keep cool until ready to use)

Recipe for popcorn ball "cement:"

237 ml (1 cup) granulated sugar

237 ml (1 cup) brown sugar (packed)

178 ml (3/4 cup) light corn syrup

178 ml (3/4 cup) water

5 ml (1 tsp.) vinegar

2 sticks of margarine or butter

Combine all ingredients except butter in a sauce pan. Heat to boiling over medium heat while stirring frequently until a small drop of the mixture dropped in ice water forms a soft ball (127°C or 260 °F on a candy thermometer). Reduce heat to low and stir in butter until melted.

The Author

Christy D. McGee is a clinical instructor at the University of Louisville.

7 Play-Doh Stratigraphy

BY LEE A. KROHN

Focus:
Pressure from the weight of the overlying sediment compacts sediment layers. This is simulated in the activity by pressing down on layers of Play-Doh with the palm of your hand. The process of compaction, along with other physical and chemical processes, eventually transforms sediment layers into rock. Tectonic pressures may further deform the rock layers into structures such as upfolds and downfolds which are readily seen along road cuts and at rock outcrops.

Challenge:
How can you simulate deformation of rock layers by pressure and folding? Do your results look similar to any rock formations you have seen? What are the limitations of this model?

Time: 50 minutes

Procedure:
1. Roll out half a can of each color of Play-Doh to a thickness of between 5 mm and 10 mm.

2. Layer the four rolled-out sheets of Play-Doh one on top of the other. Seashells or other objects could be placed into one or more layers to simulate fossils. These layers represent layers of sediments, deposited sequentially over a period of time.

3. Now, with the palm of your hand, press down hard on these layers to simulate deformation by pressure, and then examine the deformation. Be sure to point out that in actual rock formation other processes besides pressure are needed to transform sediment layers into rock.

4. Now, use both hands to compress the layers inward, simulating tectonic plates colliding.

Materials and Equipment:
Each group of students will need:

Waxed paper or other smooth, non-stick working surface

A package (four cans) of different colors of Play-Doh

A rolling pin

A butter knife

Seashells or other objects to simulate fossils (optional)

Q How could these materials be used to simulate folding of the Earth's crust?

Q Why are the layers deformed more in some areas than others?

5. Examine the layers on the outside, then slice through the layers at different angles with the knife to see what it looks like from different perspectives.

Further Challenges:

Can the students draw parallels from the above model to outcrops or road cuts they might have seen? How do geologists make interpretations from what little they can see on the surface? Try interpreting the sequence of deposition and deformation from another Play-Doh model made ahead of time.

Q Which aspects of these simulations of geologic events are realistic? Are unrealistic? In other words, what are the limitations of this model?

The Author

Lee A. Krohn was head of the science department at The Greenwood School in Putney, VT and is currently planning director for the town of Manchester, VT.

8 Collect and Compare

BY ROSE WEST

Focus:
This activity focuses on the skills of identifying rocks and comparing those rocks to buildings and other structures.

Challenge:
What types of rocks are used in different manufactured structures (buildings, monuments, roads, bridges) in your area? How are the rocks used?

Time: Five class periods

Procedure:

1. Examine samples of sandstone, granite, and other rocks used to build various structures. These are available from masonry contractors, landscaping stores, or makers of cemetery headstones. Note distinguishing characteristics such as crystal size and color.

2. Make a data table in the class log book with two columns. Label one column "Type of Rock" and the other column "Structures Made of This Rock."

3. Now, go on a field trip around your area, and compare the rock samples students examined in step 1 to any buildings, bridges, or other rock structures such as landscaping borders or gravestones. It may be helpful on the field trip to take representative samples of each rock type examined in class.

4. For each structure, enter in the log book what the structure is and which rock or rocks from the classroom samples have been used in the structure. Students may take turns making entries in the appropriate columns of the log book.

As an option, the students could take turns photographing or sketching the structures that match the rock samples. Another possibility is to buy postcards of the structures, if they are available.

5. Bring the log book, rock samples, photos, and sketches back to the classroom. Use a rock identification book and student input to identify the rock samples.

6. Finally, make a display by gluing to a poster board the rocks and, next to them, the pictures, sketches, or written descriptions of the corresponding structures made of those rocks. Label each entry and put up the poster in the classroom as a continuing lesson.

Materials and Equipment:
The whole class will need:
Samples of rocks used in buildings (sandstones, limestones, granites, etc.)
Rock identification book
Log book
Camera (optional)
Glue
Poster board

Further Challenges:

Collect rocks from the local area, and search for places where those types of rocks were used, such as garden walks or stone walls. Or see how many different types of rocks are used in and around home and school.

References:

Barnes-Swarney, Patricia L. (1991). *Born of heat and pressure.* Hillside, NJ: Enslow Publishers.

Hamilton, W. R., Woodley, A. R., and Bishop, A. C. (1974). *Minerals, rocks, and fossils.* New York: Larousse and Co.

Henrod, Lorraine. (1972). *The rock hunters.* New York: G. P. Putnam's Sons.

Pine, Tillie S. and Levine, Joseph. (1967). *Rocks and how we use them.* New York: McGraw-Hill Book Co.

Pough, Fredrick, H. (1960). *A field guide to rocks and minerals.* Boston: Houghton Mifflin.

Tilling, Robert. (1991). *Born of fire.* Hillside, NJ: Enslow Publishers.

Zim, Herbert, and Schaffer, Paul R. (1957). *Rocks and minerals, a golden nature guide.* New York: Golden Press.

The Author

Rose West teaches science at Huth Upper Grade Center in Matteson, IL.

9 Smaller than Your Pinky, Larger than Your Fist

Square, Round, Hard—What Is a Rock?

BY BETTY B. GRAHAM

Materials and Equipment:

Each student will need:

2 scavenger hunt lists (school yard hunt and take-home assignment)

A lunch-sized paper sack

2 plain sheets of paper (14 cm x 20 cm or larger)

A hand lens

Water (optional; for washing rocks)

A paper towel

Focus:

Rocks are fascinating tactile objects that compel children and adults to hold, feel, rub, and otherwise examine them. Studying rocks is a beginning for inquiry and investigation which leads to different discoveries and, in turn, to new investigations.

These discoveries may be on different levels. They might be as basic as discovering a color never seen before—an Earth red, shiny black, or copper blue-green. Or they might be as advanced as discovering that many rocks indigenous to the area are sedimentary and contain fossils. This, in turn, might lead to investigating the types of fossils in the sedimentary rock. Other discoveries might include the number of holes in a rock, the variability in the size of those holes, the number of different materials making up the rock, or the nature of those materials—whether they are small specks, crystals, or large chunks.

Challenge:

What is a rock? How are rocks similar? How are they different? Can big rocks become small rocks, sand, and dirt? Can sand and dirt become rock?

Time: 40 minutes

Procedure:

1. Begin class by asking the students what a rock is and how a rock is different from a seed, a sugar cube, a fingernail, a button, a paper clip, or other objects. Ask them if a rock is a living thing.

2. As a class, prepare two scavenger hunt lists which might include categories such as color, size, shape, or texture. See the sample list at the end of this activity. The Class Time list introduces students to making choices and decisions and to recognizing shapes, sizes, and color. The Take Home list expands on the Class Time list; however, a student may bring a rock that is a "perfect this-or-that" from the other list.

3. Explain that the object of a scavenger hunt is to search for and find as many things on a list as possible. Give each student a paper sack and a Class Time scavenger hunt list,

and proceed outside to try to find as many rocks as possible that match the categories on the list. Urge students to choose their rocks carefully to fit the descriptions.

4. After 15 or 20 minutes, return to the classroom and place each student's collection of rocks on a paper towel with the student's name on it. (Washing and drying of the rocks is optional.) The number and maturity of the students should determine how to share.

5. Return the paper sacks, and distribute the Take Home scavenger hunt list for a home assignment. The students may borrow some rocks from rock collectors, but they should NOT buy rocks to fit the list. Point out that they might find only a few of the rocks on the list, but the looking, searching, and observing is the challenge.

6. During the next class, repeat step 4 for the rocks brought from home, and have each student examine his or her rocks from the two scavenger hunts with a hand lens.

7. Have each student show and tell why some of their rocks were chosen to represent particular categories.

8. Then, have students draw one of their rocks from each scavenger hunt on a sheet of the plain paper. Observations may also be in words. Title the paper "My Observations (or What I discover with all my senses.)"

Further Challenges:

Compare the rocks collected for the scavenger hunts to large rock specimens in the classroom and discover the kinds of rocks and/or minerals that have been collected. Can they match their rocks to components of the large specimens? For example, a small pink rock may be the same as the feldspar in a chunk of granite.

Have them research the kinds of rocks found around the school. Are they naturally occurring in your area? If not, how and why were the rocks moved to the school grounds?

References:

Barnes-Swarney, Patricia L. (1991). *Born of heat and pressure.* Hillside, NJ: Enslow Publishers.

Hamilton, W. R., Woodley, A. R., and Bishop, A. C. (1974). *Minerals, rocks, and fossils.* New York: Larousse and Co.

Henrod, Lorraine. (1972). *The rock hunters.* New York: G. P. Putnam's Sons.

Pine, Tillie S. and Levine, Joseph. (1967). *Rocks and how we*

The Author

Betty B. Graham teaches science at E. J. Martinez Elementary School in Santa Fe, NM.

use them. New York: McGraw-Hill Book Co.

Pough, Fredrick, H. (1960). *A field guide to rocks and minerals.* Boston: Houghton Mifflin.

Tilling, Robert. (1991). *Born of fire.* Hillside, NJ: Enslow Publishers.

Zim, Herbert, and Schaffer, Paul R. (1957). *Rocks and minerals, a golden nature guide.* New York: Golden Press.

Class Time Scavenger Hunt List

A scavenger hunt means to look for things on a list and then collect them. Look for the rocks described below in your school yard. Bring the ones you have found when you are called.

1. A rock smaller than your pinky.

2. A rock larger than your fist.

3. A square rock.

4. A smooth rock.

5. A rock that has several different colored specks in it.

6. A very rough rock.

Observe the rocks using a hand lens.

Draw and write what you observe.

Take Home Scavenger Hunt List

Bring as many of the following rocks as you can find to class.

1. A rock that is all the same color.

2. A rock that has different colored specks in it.

3. A rock that is flat.

4. A rock that is bumpy.

5. A rock that is shiny.

6. A rock that is dull.

7. A rock that is dull but has shiny pieces in it.

Be fair—try to find your own rocks. If you know a rock collector, ask if you may bring a rock from their collection. Do not buy rocks for this activity. Do your best!

Rock and Roll

BY PAUL CAULFIELD AND ROSE WEST

Focus:
This is an opportunity for students to go beyond passively demonstrating their knowledge of Earth science by having them challenge each other intellectually. The students will develop questions based on their textbook to strengthen their questioning skills.

Challenge:
How many questions and answers about Earth science can you come up with that are challenging to your classmates?

Materials and Equipment:
The whole class will need:
A rock 'n' roll cassette tape
A tape recorder
Each student will need:
A textbook
A marker
A pen
Construction paper
White paper
Glue

Time: One class to write questions, one class to make answer sheets, then as many days as desired for the musical question part

Procedure:

1. Ask the students to come up with five questions and answers about rocks and related topics from their Earth science textbook.

2. Collect and read the questions and answers, and circle each student's best question. Avoid duplicate or similar questions. If any concepts are missing, suggest these to students who need help.

3. Return the papers along with a piece of white paper, a piece of construction paper, a marker, and glue for each student.

4. Ask the students to fold the construction paper in half. Next, have them write their best questions (the one you circled) on the plain white paper and glue this to the inside of the folded construction paper. Assign each student a number. Have them write the number on the lower right corner on the outside cover of the construction paper, and then draw a design on the cover.

5. After all question sheets are done, ask each student to number a piece of notebook paper from one to the total number of questions prepared by the class. To add a little more fun to the activity, you might add one or two of your own silly questions.

6. Direct the students to move one desk to the right, leaving their designed question on the desk. Students who sit at the far right desk in a given row move to the first desk at the far left of the row directly behind them.

7. Play music for one minute during which time the students are to open the paper and answer the question. When the music stops they are to move to the next question, again leaving it behind for the next student. Allow 15 seconds to move to the next desk.

8. When the rotation is complete, collect and correct the papers. You may want to laminate the question sheets for future use.

Further Challenges:

Students can create their own set of ten or more questions, or repeat the activity for another subject; for example, Rocket and Roll, Shake and Quake, or Whispering Weather.

The Authors

Paul Caulfield and Rose West teach science at Huth Upper Grade Center in Matteson, IL.

Weathering or Not?

BY SHIRLEY G. KEY

Focus:
Weathering is the set of processes that changes the size and/or the composition of rocks and other materials and thereby changes the environment. The two general types of weathering are chemical and mechanical. They usually occur very slowly and people usually don't notice them happening.

Chemical weathering processes change the composition of rocks and other materials. Some visible examples of chemical weathering are rusting and discoloration of buildings or other structures.

Mechanical or physical weathering processes, on the other hand, do not change the composition of the rocks and other materials but break them into smaller pieces. Some examples of mechanical weathering include the pounding surf and water freezing within cracks of a rock and expanding.

Challenge:
What is the difference between mechanical weathering and chemical weathering? Can chemical and mechanical weathering change your school's appearance? How can weathering improve the environment?

Time: 3 days

Procedure:
1. To simulate chemical weathering, begin by pouring some seltzer water into a clear plastic cup.

2. Drop a piece of marble or limestone into the cup and observe what happens.

3. Now, to simulate mechanical weathering, place two pieces of sandstone or limestone in a plastic container that is half filled with water, and secure the lid.

4. Vigorously shake the container for one minute, then take the rock out and examine it.

5. Pour the water from the container into a clear plastic cup, and examine the sediment remaining in the container.

6. Compare chemical weathering in the first part of the activity with mechanical weathering in the second.

7. Now, divide the class into two groups, giving each group a camera and a log book, and go outside to look for signs of weathering. One group will keep a log of sites of mechanical weathering and the other will keep a log of sites of chemical

Materials and Equipment
The whole class will need:
Two 2-L bottles of seltzer water
2 cameras and black and white film (12 exposures for each camera)
2 log books
Each group of students will need:
2 clear plastic cups
A small piece of marble or limestone
A plastic container with lid (empty margarine tub)
2 small pieces of sandstone or limestone
Water

Q How is the seltzer water affecting the stone? What would happen to a statue or building if rain was made of seltzer water?

Q How has the piece of sandstone changed? Where did the sediment come from? What was happening to the sandstone while the container was being shaken?

Q What weathering processes are at work at each site? Could both mechanical and chemical weathering occur at the same site? What changes are likely to occur at these sites over time (six months, five years, 100 years)? Are these changes inevitable?

weathering. Places to find signs of weathering include the corners of a building, sidewalks, gutters, streets, potholes, fences, bayous, ditches, and old trees. Some sites may exhibit both types of weathering and so would be photographed by both groups.

8. Take pictures of the sites after deciding whether it exhibits chemical or mechanical weathering. Date the picture, record the type of weathering, and discuss what might have caused the weathering.

Further Challenge:

Keep the log books, and for a designated period of time (such as a month, three months, or a school year), take pictures of the chemical and mechanical weathering once a week at these same sites, noting any changes. On the designated final day, ask the two groups to explain all changes they have observed. Conclude by discussing what weathering is and how it can affect the environment. Are there ways to slow down the processes of weathering?

You may want to use a stronger acid such as vinegar or lemon juice in the chemical weathering activity to better illustrate the effects of acid rain. Place such materials as metal, wood, and glass into water as well as the more acidic substances and compare the observations. Are vinegar and lemon juice more destructive than just water? If it were to rain lemon juice, would buildings and rocks weather faster or slower than if the rain were just water?

The Author

Shirley Gholston Key was a science teacher at Christa McAuliffe Middle School in Houston, TX and is currently a doctoral student at the University of Houston.

Dirt Cake

BY KAREN K. LIND

Focus:
Soil is made up of layers of rock fragments (sediment) and organic material (humus). There are different combinations of these layers, but one principle holds true: The bottom layer was laid down first. A layering of cookies, cream cheese, and pudding can simulate this principle.

Challenge:
Have you ever made dirt cake? Can foods be mixed together to look like layers of Earth?

Time: One class period to make and 2 to 3 hours to refrigerate

Procedure:

1. Crush one large package of chocolate sandwich cookies in a blender, or for more student involvement, put the cookies in a food storage bags and crush them with a rolling pin. Now add Gummy Worms to the crushed cookies or "soil mixture."

2. Mix the cream cheese and powdered sugar together. Depending on the ages of the students, you may want to do the measuring yourself or direct one or two students to carefully measure out the necessary ingredients (a good introduction to measurement).

3. Mix the instant pudding with the milk, add this mixture to the cream cheese mixture, and stir in the whipped topping.

Materials and Equipment:
The whole class will need:

A large package of chocolate sandwich cookies

158 ml (2/3 cup) of powdered sugar

Two 226-g (8-oz.) packages of cream cheese

158 ml (2/3 cup) milk

2 packages of french vanilla instant pudding

Gummy Worms

353 ml (12 oz.) whipped topping

Silk or plastic flowers or ferns

A plastic flower pot

A clean garden trowel

A blender or a food storage bag and rolling pin

An electric mixer

Two large mixing bowls

Measuring cups

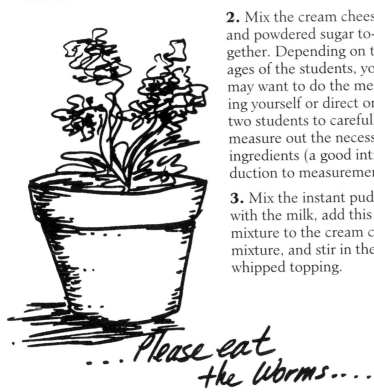

...Please eat the worms....

Q Which layer was potted first? Potted last?

The Author

Karen K. Lind is an assistant professor in the school of education at the Louisville in Louisville, KY.

4. Time to pot your soil. Start with a layer of crushed chocolate sandwich cookies in a clean plastic flower pot. Then alternate layers of the cream cheese and pudding mixture with crushed chocolate sandwich cookies until the pot is full. Be sure to make the last layer cookies so it looks like dirt on top.

5. Refrigerate your simulated Earth layers for 2 to 3 hours.

6. Before serving, add the plastic or silk flowers. Add a sign: "Dirt Cake" or "Please Eat the Worms," and serve with the clean garden trowel. Point out that the layers are still distinct and are in the same order as they were placed in the pot.

Further Challenges:

Can you simulate other natural phenomena? How might we make Mississippi Mud Pie?

References:

Strange, Johanna. (1988). *CESI make and take.* Louisville, KY: Association for Progress in Science.

Shake and Sieve!

BY SUE DALE TUNNICLIFFE

Focus:

Soil contains water, rocks, minerals, and organic materials that plants need to grow and animals depend on either directly or indirectly for food. This makes soil necessary to our lives. As a result of the processes of weathering, soil contains particles of rock and organic materials. Students can examine these particles by sifting soil through sieves they make themselves.

Challenge:

What type of apparatus will separate particles of soil? Can you separate the largest particles of soil from the rest of the sample?

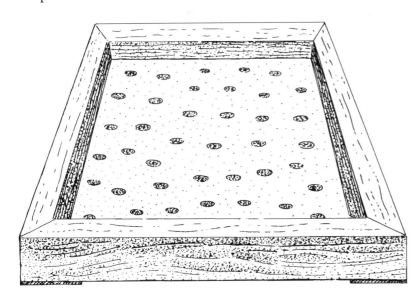

Time: Two hours

Procedure:

1. Let the students examine the soil samples with their eyes and hands.

2. Challenge students to think of ways to sort the largest particles from the rest of the soil other than by picking them out by hand. Steer their conversation and ideas to sieving. Show an example of a sieve such as a child's beach sifter, and ask the students how they would go about constructing a sieve.

3. Divide the class into groups of two or more so that the students can share and refine their ideas on how to design and build a sieve. Give each group a list of the items to

Materials and Equipment:

Each group of two or more students will need:

Several types of soil

A large paper cup

An aluminum pie tin

A sieve, which will require:

Graph paper for drawing a plan

List of possible items to construct the sieve:

4 equal lengths of thin plywood, pencils, or strengthened soda straws for the frame

Cardboard triangles to strengthen the frame corners

Wood glue, masking tape, or staples

Cardboard for a sieve mesh

A metric ruler

Scissors

A hack saw

A vice or bench hook for cutting wood

Sandpaper for finishing edges of wood

Glue or thumbtacks for attaching mesh to frame

Q Can you identify the largest particles by eye?

construct the seive and a sheet of graph paper to draw their design on. Encourage students to think through the design process by having them do the following.

• Identify aloud the task at hand. Each student should be able to explain it in his or her own words.

• In designing, think carefully about sizes and materials.

4. Have the groups work together to draw a design, complete with measurements, on the graph paper. It is helpful to draw the design full scale. Suggest notes on materials to be used and techniques needed. For example, if the frame is to be made of wood, you will also need a saw; sandpaper; a vice, bench hook, or C-clamp; a pencil for marking; and a means of joining the pieces. (Note any modifications to the design you make as you proceed. Design is an ongoing process.)

5. The following steps are from a proven sieve design. You might suggest them to any groups having particular trouble or to all groups after they have finished their designs but before they begin actual construction.

Lay the four pieces of wood over the plan, if drawn to scale. (This helps to ensure better right angles.) Glue two pieces together so that you have two right-angle sections. Glue the other two corners together when the first joints are dry.

6. In joining together the pieces of wood, the corners can be reinforced with cardboard triangles. As shown in the figure, glue the wood together, and then add cardboard triangles to the top and bottom of the joints for extra strength. You will need eight triangles for each square frame.

7. In deciding on the size of holes to make in the cardboard mesh, students can measure the particles to be sieved and then measure and cut smaller holes in the cardboard.

Remember that in designing the mesh, the cardboard area must be sufficiently larger than the inside dimensions of the frame so that it overlaps the edges and can be fastened to the frame. Encourage the students to work this out and to determine, with a small piece of wood and sample cardboard, the most effective way of securing the cardboard to the frame.

8. Once the sieves are complete, have the students sieve the different samples of soil. One student can hold the frame and gently shake it while another pours the soil through it. A third student can hold the pie tin to catch the sieved soil.

9. Evaluate the finished product.

Q Does your sieve do what it was designed to do or are further modifications needed? Could you improve the design in any way?

Further Challenges:

Try shaking the sieve more vigorously. What effect does shaking the sieve have on the amount of time needed to sift the sample? Does shaking affect the size of particles that pass through the sieve? How can you make this a fair test by measuring the effect of the shaking so that it is the same each time? Can you devise a way of comparing the largest soil particles in different soil samples?

You may want to try a variety of ready-made meshes such as window screening, loose-weave fabrics, or small-holed garden mesh and compare their sieving capabilities.

The Author

Sue Dale Tunnicliffe is Head of Education of the Zoological Society of London.

14 Soil Profiles

BY DAVID R. STRONCK

Focus:

Several characteristics of soil including texture, color, and composition change the deeper one digs below the surface. Usually decaying matter and living plants and animals can be found on the surface and in the upper layers of soil.

The thickness of this organic layer is variable, but the amount of biological material will lessen with depth. In rain forests and deserts, there is often only a thin layer (or no layer at all) of biological material near the surface. But some soils, like those in rich farmlands, have a very deep organic layer. Perhaps the most curious soils are in bogs, where top layers are so rich in plant material that they burn easily when dry.

Challenge:

What changes do you notice in soil characteristics at different depths below the surface of your school yard? Are there living organisms in each layer? Are there dead organisms in each layer? Do the type and number of each organism change with depth?

Time: 50 minutes

Procedure:

1. Go out to the edge of the school yard or playground and select a site for digging. Before digging, examine the surface and note any living plants or decaying leaves.

2. Start digging, and notice that the top layer of soil is rich in living organisms or plants and animals that have recently died, but as the students continue to dig deeper, the amount of biological material diminishes. Also note if the top layer is dark and the lower layers become more pale and rocky.

If the soil shows little or no change in the top 60 cm, it may be impractical to dig deeper. If a trench is already dug or a road is cut through a hill, there may be little need to dig a hole.

3. Describe other soil profiles such as those of a rain forest or farmland and ask students to compare their profiles to the descriptions of these different soils.

4. Take samples of the soil in a vertical column from the surface to the bottom of the hole, placing each sample in a sandwich bag. Note the depth at which the samples are taken. Taking samples from a recent road cut, streambed, or construction site could provide a more complete profile of

Materials and Equipment:

Each student will need:

A shovel or hand trowel

A hand lens (optional)

A cardboard box at least 1 m in circumference

6 sandwich bags

✚Safety Note:

The Further Challenges activity is best done as a demonstration for younger students. Have baking soda nearby to extinguish any samples that catch fire.

Q Can you predict what the soil will look like as you dig deeper and deeper below the surface of the school yard?

Q What traces of plants or animals can be found below the surface of the soil? Are there distinct layers of soil types below the surface? What are the characteristics of each layer?

the soil layers in your area.

5. Before returning to the classroom, carefully fill in the holes and restore each digging site to its original condition.

6. Break apart a cardboard box so that it makes a sheet about 1 m long. Organize the samples on the cardboard sheet or a wooden board, labeling the layers according to the depth they were taken from, and describe some of the characteristics of each sample in writing.

Further Challenges:

Students can test the different layers for ability to absorb water. Take equal quantities of soil from each layer, and pour equal amounts of water into the samples. Which sample absorbed the water most quickly?

Instead of dousing the samples in water, try the opposite and heat them up. Cook individual samples in a pan on a stove or hotplate. Which samples give off smoke? Which catch fire and burn? Which seem only to become dry from the cooking?

The Author

David R. Stronck is a professor of teacher education at California State University in Hayward and is currently NSTA Research Division Director.

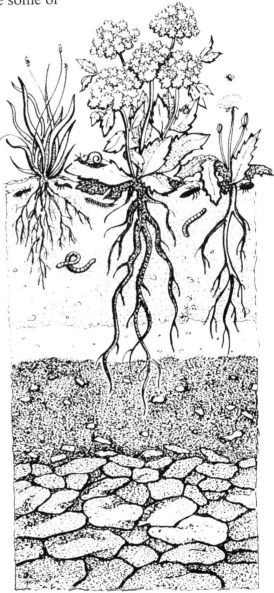

Water, Stones, & Fossil Bones

15 Big and Little Sand

BY LLOYD H. BARROW

Focus:
Sand grains come in many different sizes. Recognizing this feature by examining various grades of sandpaper presents the concept of classification based on a physical characteristic.

Challenge:
Which type of sandpaper has the largest-sized sand grains? What characteristics could you use to classify sand grains?

Time: 40 minutes

Procedure:
1. Ask the class what sand is and where it comes from. Then, to further pique their curiosity about the nature of sand, direct students to use the hand lens to examine the coarse sandpaper. Have them take note of the size of the sand particles and their corners. On sand paper that has been used, the corners will have been worn off.

2. Gently rub the coarse sandpaper once across one of the plastic soda bottles, then using the hand lens, examine the scratches made on the bottle. Repeat, using the fine sandpaper on the other soda bottle.

3. Now use the hand lens to examine the sand from a sandbox or beach, and notice the size, color, and shape of the sand grains.

4. Put two handfuls of this sand in some panty hose or on some other screening material and gently shake it over an empty shoe box.

Further Challenges:
Give each student or group a piece of medium sandpaper. Have them predict how this grade of sandpaper will scratch a plastic soda bottle compared to the other two grades. Then direct them to test their predictions by scratching three plastic soda bottles with the three grades of sandpaper.

Ask how many sand particles there are in a square centimeter for the fine, medium, and coarse sandpaper, and then count them to find out.

References:
Adapted from a forthcoming book *Adventures with rocks and minerals* from Enslow Publishers, Box 77, Hillside, NJ 07205.

Materials and Equipment:
Each group of up to four students will need:

Two types of sandpaper (fine and coarse)

A hand lens

Sand from a beach or sandbox

Old panty hose or screening

Two 2-L plastic bottles

A shoe box

Q Which type of sandpaper made the widest scratches on the bottles? Which type made the most scratches?

Q How do the sand particles that stayed in the panty hose compare with those that passed through the panty hose?

The Author
Lloyd H. Barrow is a professor of science education at the University of Missouri in Columbia and is currently NSTA District VIII Director.

Mini Landfills

BY MILDRED MOSEMAN

Focus:
Some communities dispose of their garbage by burying it under a layer of soil. This is called a landfill. Some kinds of garbage decompose more slowly than others. Some materials change very little after 10 or 100 years or even, as in the case of aluminum, 1,000 years. Synthetic textiles such as nylon, rayon, and synthetic leather; synthetic insecticides and fungicides; and plastics cannot by decomposed by microorganisms.

One of the problems with landfills is that eventually the surface may settle and slowly sink. One corner of a three-year-old convention center in Sioux City, Iowa is already beginning to sink, and in Los Angeles a landfill 30 m deep sank almost 2 m in three years.

Challenge:
Create your own miniature landfill. Can you estimate how long it would take different types of garbage to decompose by observing what happens in your miniature landfill?

Time: Four weeks

Procedure:
1. Put a layer of loamy soil about 2 cm deep in the cup, and then place two or three pieces of garbage against the sides of the cup so they are visible from the outside.

2. Put in another layer of soil, and place a second layer of garbage in the same manner as the first. Continue layering until the cup is almost full.

3. Firmly pack the top layer of soil, leaving approximately 2 cm of air space above it.

4. Spray the miniature landfill until it is moist (but not puddled or water-logged), and then cover the cup with plastic wrap and secure it with a rubber band.

5. Keep the landfill moist, spraying it when the condensation on the plastic wrap begins to disappear.

Materials and Equipment:

Each group of students will need:

A large, transparent plastic cup, big enough to place your hand in

Plastic wrap

A rubber band

Loamy soil

Garbage such as nut shells, potato peels, cereal, apple cores, orange peels, paper napkins, plastic from a garbage bag, newspaper, or aluminum foil

A spray bottle of water

Q Can you see any organisms other than the mold inside the landfill? Are the spaces you see, where the garbage used to be, really empty? What happened to the bulk of the garbage materials that originally occupied these spaces? Where did the materials go? Why does the surface of a landfill sink? What would happen to houses built on a landfill?

The Author

Mildred Moseman, currently retired, taught at the Lincoln School in Sioux City, IA.

6. Over the next several weeks, record how rapidly the different garbage items decompose. Students will begin to see various types of molds appear in their landfills within a few days. As items decompose, they will leave empty spaces.

Further Challenges:

A landfill 2.13 m deep and 0.4 hectares in area can hold one year's garbage from 10,000 people (1 hectare = 10,118 m^2). Find out the population of your community, and calculate how many hectares of land would be needed to dispose of all your community's garbage produced over a 10-year period. What if a landfill can be 4.26 m deep—how many hectares of land are needed to hold one year's garbage from 10,000 people? How many hectares would your community need to dispose of its garbage produced over a 10-year period?

References:

Schatz, Albert, & Schatz, Vivian (1971). *Teaching science with garbage.* Emmaus, PA: Rodale Press.

A Model of the Earth's Crust

BY GERALD WM. FOSTER

Focus:

One of the simplest ways to obtain a soil profile is to use a hollow bored auger. The auger is pushed directly down into the soil and pulled up by a horizontal handle. The soil layers in the auger are in the same position as they were in the ground. In addition the thickness of the layers can be measured as well as their composition examined.

In exploring the layers of the Earth's crust, the core sample may not be adequate or practical to obtain. Modern technology now offers geologists more sophisticated ways to explore the layers of the Earth's crust. For example, sound waves or radio waves can be sent through the layers to determine their thickness, depth, and composition. Geologists, and scientists in general, often use more than one method to supplement and confirm information they already have, a very useful approach when attempting to construct a model of some geologic formation whether it is a soil profile or a model of the Earth's crust.

Materials and Equipment:

Each group of four students will need:

A shoe box with a lid

Several different soil types such as clay, potting soil, and sand

Metal rods (large knitting needles)

Large, sturdy, clear plastic straws

A metric ruler

Drawing paper

Challenge:

By probing with a metal rod, can you determine the order of different soil layers? How does your description compare to a core sample taken from the model soil profiles?

Time: 90 minutes

Procedure:

1. Prior to class, prepare a shoe box for each group that contains layers of

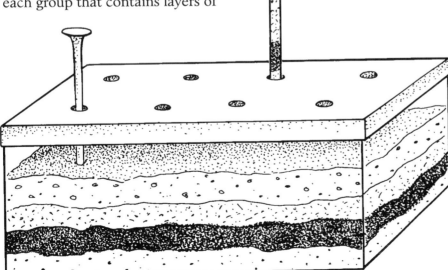

Water, Stones, & Fossil Bones

the different types of soil. Each layer should be at least 2 cm thick or thicker, and each shoe box must have the same profile.

2. Make eight evenly spaced holes in the shoe box lids big enough for the metal rod or the straw to fit through (whichever is larger), but small enough so that students cannot see the samples.

3. In class, show the students samples of the different soils layered in the shoe boxes. Explain that these soils make up the different layers and their task is to determine the order of the layers and describe them after probing the layers with a metal rod.

Q Does the rod move through some layers more easily than others? Are the soil particles within a layer all one size? How can you tell if a layer has rocky fragments in it?

4. Have students carefully probe the soil layers with a metal rod and note how easy it is to push the rod through the layers as well as friction from soil particles. Based on this information, have each group write down a proposal for the sequence of the soil layers. The next step will test their proposal.

5. Using the clear plastic straws, obtain a core sample of the soil layers inside the box. Students may have to moisten the soil layers in order to come up with good core samples in the straws.

6. Have the groups describe to one another their soil samples, and then draw a scientific model depicting the sequence of soils as well as the thicknesses of the various layers. By developing their own models and discussing them with each other, students will have an opportunity to understand that scientists often arrive at different models even though they may be using the same evidence.

Further Challenges:

Take pictures of soil profiles where there is construction and bring them to class. How thick are the layers relative to each other? What is the composition of each layer? What is the orientation of the layers with respect to each other and to ground level?

The Author

Gerald Wm. Foster, Ph.D., is an associate professor of Science Education at DePaul University in Chicago.

Students could also research how scientists develop models of the Earth's crust and the methods they use. Or have a soil scientist, perhaps from a nearby college or university or from the county extension service, come to class to explain the soil characteristics in the neighborhood or school yard.

Model of a Seismograph

BY MUHAMMAD HANIF

Focus:

Seismologists record and measure the motion of earthquake waves with an instrument called a seismograph which responds to the motion of the ground surface beneath the instrument. A seismogram, the zigzag line made by a seismograph, is a record of the varying amplitudes of seismic waves, and from it scientists can estimate the energy that was released by an earthquake.

A seismograph is based on the principle of a pendulum. Because of its inertia, the heavy mass at the end of the pendulum will remain still while the instrument moves underneath the pen at the time of the earthquake.

Challenge:

How can you measure a table-top earthquake? Can you construct a model seismograph? How does the seismograph record an earthquake? On a seismogram, how are the amplitudes of seismic waves represented?

Time: Two to three class periods

Procedure:

1. Nail together a wooden bracket, as shown in the figure on the next page, from the lengths of 2 x 4 wood.

2. Hammer four nails into the top side of the base so that all the nails are exactly 2 cm tall and together form the corners of a 15-cm x 5.5-cm rectangle as shown below. This will serve as a rest for the barrel of the seismograph (the plastic bottle).

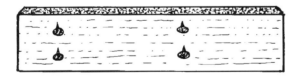

3. Wrap a sheet of white paper around the empty plastic bottle, and tape it in place. Then turn the bottle on its side and set it on the four nails at the base of the bracket. Make sure the nails are not touching the paper so they will not tear it when the bottle is rotated. As an option, fill the bottle half full with water so it will rotate more smoothly on the nails.

4. Connect the 1-kg mass to the end of the thread to serve as a mass that will provide inertia. Attach the felt tip pen to the mass with rubber bands.

5. Screw a cup hook into the underside of the top arm of the

Materials and Equipment:

Each seismograph model will require:

Three pieces of 2 x 4 wood (one 61 cm long, two 23 cm long)

Eight 6.3 cm (2 1/2") nails

A cup hook

String

A 1-kg mass

A felt-tip pen

A 1-L plastic bottle

6 sheets of plain white paper (12.7 cm x 28 cm)

Tape

A metric ruler

Rubber bands

bracket from which the thread with the mass and pen will hang. Suspend the string from the hook, adjusting its length so that the tip of the pen barely touches the very top of the bottle and the sheet of paper covering the bottle.

6. Place your model seismograph on a table and steady the 1-kg mass so it is not swinging. Rotate the bottle and observe the shape of the line traced on the paper.

7. Replace this piece of paper with a new sheet, and steady the mass again.

Q Do you notice any changes in the traced line?

8. Now, shake the table back and forth while continuing to rotate the bottle. Make sure not to lift the table since the seismograph is not designed to record vertical movement. Also, the string should not start swinging like a pendulum, only the frame holding the bottle should move. Compare the lines on both sheets of paper.

Q Are all earthquakes and seismic waves the same?

9. Put another piece of clean paper on the bottle. Now, increase the intensity of the table's motion. Students will see from the results recorded by the seismograph that the table was sometimes shaken harder than at other times. The increase in the amplitude of the wave from the zero position indicates an increase in motion and in readings on the Richter scale. You might also create an earthquake by jumping up and down on the floor.

Q What would cause some earthquakes to be more violent than others?

10. Study the seismograms and put them in order from least amplitude to greatest. Explain that earthquakes release different amounts of built-up energy.

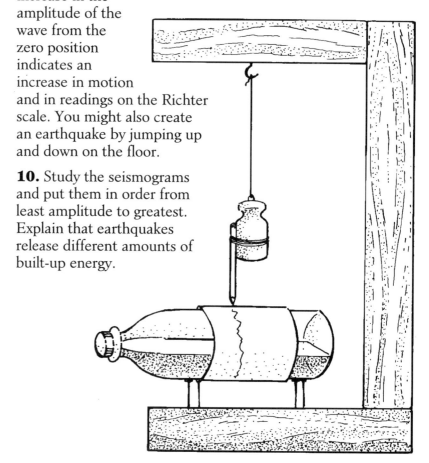

Further Challenges:

Students may want to research seismograms of well-known earthquakes such as the September 15, 1985, earthquake in Mexico City. Contact the United States Geological Survey, P.O. Box 25425, Denver Federal Center, Denver, CO 80225. Ask for copies of seismograms of any four or five major earthquakes.

References:

Calder, N. (1972). *The restless Earth: A report on the new geology.* New York: Viking Press.

Cazean, C. J. (1977). Earthquake. *Instructor, 86*(6), 76–78.

Eiby, G. A. (1980). *Earthquakes.* New York: Van Nostrand Reinhold.

Hanif, Muhammad (1990). As the Earth quakes...What happens. *Science & Children, 27*(4), 36–39.

The Author

Muhammed Hanif is an associate professor in the School of Education at the University of Louisville.

19 Songs of Earth Science

BY ROSE WEST

Focus:
Topics such as plate tectonics, earthquakes, and volcanoes become themes for songs written by students. This activity is a creative approach to expanding and strengthening students' knowledge of Earth science.

Challenge:
Can students use what they know about plate tectonics, earthquakes, and volcanoes to write songs about these different topics of Earth science?

Time: Three class periods

Procedure:
1. Write the names of the song topics on separate pieces of paper, for example, earthquake, volcano, and tsunami (one topic per group of students). Place the pieces of paper in a hat, and ask one student from each group to select the topic for their group.

2. Each group is to create a song that includes as many facts as possible about their topic. The background music for their songs can be anything from rap to rock to an original melody.

3. For their topic, each group will create a poster or model to illustrate their song. Allow class time for songwriting. The poster or model can be assigned as homework or it can be done in class as well.

4. Each group will present their composition to the class. Students could wear costumes for these concerts, use props and musical instruments, or create their own backup sound with the doo-wop rap noise made with the mouth.

Further Challenges:
The groups could take their act on the road performing their songs for other classes. Or write raps for other topics in Earth science such as weather, glaciers, rock formation, or for other subjects entirely.

Materials and Equipment:

Each group of three to five students will need:

Earth science books

Paper

A pen

Posterboard

A marker

A musical instrument (optional)

The Author

Rose West teaches science at Huth Upper Grade Center in Matteson, IL.

"Rap" Sheet

The Volcanoes

Chorus—We're the volcano erupting crew. Our lava's pouring out and it's coming to get you.

Verse—1st person: I'm a hot-spot volcano. I'm close to the ground. When my lava comes out, you better get out of town.

2nd person: I'm a continental volcano. I live on the ground. Everybody knows I don't mess around.

I can burn, I can sizzle any time of the day. When I erupt you better get out of the way.

(Repeat chorus)

3rd person: I'm a magma pool. I come out the spout. I'm really hot, so you better look out.

4th person: I'm an island volcano. I'm in the water, and when my lava comes out, you'd better holler.

(Repeat chorus)

Fade out chanting: VOLCANO. . . .VOLCANO. . . .VOLCANO

Mountain Mash

Sung to the tune of "The Monster Mash," one student can sing the verses, and everyone sings the chorus. Everyone not singing a solo should sing "Wha-ooo dit dit dit" softly as background. A recording of "The Monster Mash" will help see how this can be accomplished.

Verse—I was working outside late one night, when my eyes beheld an amazing sight. Two huge plates began to rise, and suddenly to my surprise,

Chorus—They did the mash—they did the Mountain Mash.

It was a smash—when those two plates crashed.

It was a bash—there was soot and ash.

They did the mash—they did the Mountain Mash.

The magma began to rise, but didn't break through to my surprise.

Then a miracle was performed, and a dome mountain was formed.

(Repeat chorus)

Just when I though it was over and done, two more plates collided one on one.

Then the land was forced up and a mountain made. It was a folded mountain for goodness sake.

(Repeat chorus)

Then it happened once more. Hot crust moved up, but to my confusion,

the other side moved down; it was a fault block illusion.

(Repeat chorus)

Then right and left, things were shaking and up and down, things were quaking.

But mother nature saved the day, and no more mountains

came my way.

(Repeat chorus)

Plate Tectonic Theory

I move real slow, not too quick. I'm what moves the continents. I'm always moving.

I never get weary 'cause—I'm just the plate tectonic theory.

Sea-floor spreading is part of my game. Plate tectonics is my name.

I'm sea-floor spreading, new rocks are pushed up.

From cracks in the sea floor. Yup! Yup!

I form mountains and never get weary.

'Cause I'm just the plate tectonic theory.

Tsunami

There once was a volcano who lived in the ocean.

One day it made a big explosion.

This explosion made a tsunami which was very bad,

and many people lost the homes that they had.

Just after dawn, when the tsunami was gone, everyone came to see the damage.

Then they found there was nowhere to go because they had no dough.

Now that's what a tsunami does, and if you want more, give me a buzz.

Earth Science—Discovery and Investigation

Discovery is an important word in science. Discovery is the result of asking questions. Where do birds sleep? How big are raindrops? What kind of rock do I have? Discovery begins with curiosity—an interest in how the world works, a way of looking at things, and a willingness to test ideas. All children are curious. They are open to experience, anxious to test their theories, intuition, and imagination.

Strategies for leading students to discover usually begin with questions posed by the teacher or students. After the initial question is raised, students usually predict what they think will happen. For example, in the activity "How Low Does It Go?" students address the question "During a rainless period, how deep do you have to dig to find moist soil?" Students then select a study area, set criteria, dig, measure, and graph the moisture depth over a period of time. To help students consider variables that might affect the outcome of investigations, guide them with questions: What are we trying to find out? What shall we change?

The processes of science we teach [children] to use and the hands-on opportunities we provide equip them with useful ways to explore with an open mind the infinite variety of their world.

20 How Low Does It Go?

BY SUSAN M. JOHNSON

Focus:

What happens to rainwater is of major importance to us, whether it is running into surface bodies of water, seeping into underground spaces, evaporating, or staying in the soil. Recurring periods of drought in areas all around the world highlight our dependence on water in the soil.

This activity is best done during a dry spell, or at least when the surface of the soil in the school yard is dry.

Challenge:

During a rainless period, how deep do you have to dig to find moist soil? Graph this depth throughout the dry spell. If you water some of the dry soil surface, how deep does the water penetrate?

Time: 90 minutes

Procedure:

1. On the large piece of paper prepare a rough map of the study area that shows the approximate size of the area and any landmarks (trees, playground equipment, playing fields, etc.).

2. To get comparable results from each group, decide on a criterion for calling the soil wet. For example, to be called wet, soil may have to leave a wet mark when it is squeezed between a piece of paper towel, or it may have to clump together.

3. Have half of the groups pick spots in the yard where they expect to dig deep to find wet soil. The other half will try to find spots where they expect to find wetness immediately below the surface. Before digging, have the groups estimate how many centimeters down they will have to dig to find wet soil.

At each spot, dig a hole, measure the depth at which the soil is wet, and record the depth on the prepared map of the study area. Be sure the students understand that they must carefully restore each digging site to its original condition.

In general, evaporation will be faster from soil exposed most directly to the sun and wind, and from soil where there is the minimal decaying material.

4. As long as the dry spell continues, take depth measurements every other day near the original holes, and graph the measurements. If interest wanes or the dry spell is long, shift

Materials and Equipment:

Each group of four students will need:

Digging tools (trowels or large spoons)

Paper towels

A metric ruler

A large piece of unlined paper, about 1 square meter

A rain gauge or shallow pan, such as a round, 9-inch cake pan

Paper

A pencil

Q Are there differences between sunny and shady spots? Between the north and south sides of the school? Between grass covered spots and bare spots? Where would you expect to find animals such as earthworms, which require moist soil?

Q Some soybean farmers plant their beans 12 cm deep so they will always be moist. Would those seeds have moisture if they were planted in your school yard? How fast would you expect evaporation to be at your site?

to once-a-week measurements.

5. Keep a rain gauge or a shallow pan on hand to measure the next rainfall. Immediately after a rain, have the students measure how deep the rain penetrated. A brief downpour can wet the surface but leave a dry layer below.

Further Challenges:

Set up an experiment to compare how well different types of soil hold water by controlling for the variables of water and sunlight. Pose the following to the students. Suppose that you were a farmer during a drought. Do you think your crops would have water in the soil longer if you had sandy soil, clay soil, or potting soil? Fill containers, such as milk cartons with the lids cut off, to the same level with the different types of soil. Add the same amount of water to each container, and set them in the sun. How long does it take for the soil to be dry to the touch? Why should the same amount of water be added to each container? Why do the containers have to be the same size and shape? Why do all the containers have to receive the same amount of sunlight?

Another challenge is to calculate how much rainfall would be necessary to reach roots 6 cm deep in dry soil. Set up a sprinkler to water a small area of the dry yard. After 1 cm of water has fallen in the rain gauge, dig to measure how far the moisture has seeped. Continue in 1-cm doses until water reaches the target depth. Ask the students: Why would it be better to water a lawn for a long time than for a short time? Why would it be better to water early in the morning than in the middle of the day?

During a drought, measure the roots of plants that are growing best, such as a clump of crabgrass, and roots of plants that are dying, such as rye grass. Which have the longer roots? Which have roots that are in moist soil? Be sure to carefully restore the divots.

The Author

Susan M. Johnson is a professor of biology at Ball State University in Muncie, IN.

21 Soil Percolating

BY MICHAEL J. DEMCHIK

Focus:

Most of the rain that falls on land soaks into the ground; the rest runs off into streams and rivers. How well rain soaks in depends on a number of factors including the slope of the land, the amount of open space between soil particles (the porosity of the soil or rock), and the soil or rock's ability to allow water to pass through (its permeability).

Porosity is governed by the shape of the particles, how tightly the particles are packed, and how many different sizes of particles there are. These characteristics, and hence the amount and rate that water soaks in, can vary among different soil layers. This activity will demonstrate the variation in the rate that water moves through different soil layers. The rate that water moves through a soil is called the percolation rate.

Challenge:

Can you predict the varying rates at which water will soak into different types of soil?

Time: 35 to 50 minutes

Procedure:

1. In the school yard, dig several soil samples with the garden trowel, putting

Materials and Equipment:

Each group of students will need:

Soil samples

A garden trowel

A plastic container for each sample to be collected (margarine tubs)

3 paper cups per soil sample

A pencil

Cotton

A graduated cylinder (50 ml or larger)

A funnel

A watch or clock with a second hand

Paper

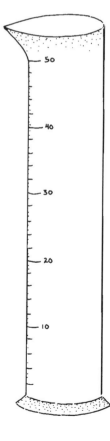

each sample into a plastic container. All of the samples should be from the same hole, but at different depths.

2. Back in the classroom, punch a hole with a pencil through the bottom of one paper cup for each soil sample collected. Then place a wad of cotton or similar material in the opening, but do not pack it tightly.

3. Fill this paper cup two-thirds full with the soil sample. With watch in hand, add 50 ml of water to it. Hold the cup over the second cup to catch the water as it flows through. Start timing when water emerges from the bottom of the cup and collect the flow for 30 seconds, or some other arbitrary amount of time that flow would be uniform.

4. Using the funnel, pour the percolated water collected during the 30 seconds into the graduated cylinder, and record the volume.

5. To determine the rate of flow through the sample, show students how to divide the amount of percolated water in milliliters by the time period in seconds. As an example, 45 ml percolated through a sample, so 45 is divided by the amount of time required for the water to pass through the soil (30 seconds): 45 ml/30 sec = 1.5 ml per second. Thus, 1.5 ml per second is the rate of flow.

Further Challenges:

Try this activity with different types of soils: sand, clay, loam, or a combination. Or select various materials such as rotted wood, dried leaves, crushed rock, or other materials of your choice that might reduce the rate of flow. Grind up each sample and place equal quantities of each in the samples tested. Make some predictions before trying these activities.

The Author

Michael J. Demchik, Ph.D., teaches chemistry and physics at Jefferson High School in Shenandoah Junction, WV.

22 River Boxes

BY DAVID R. STRONCK

Focus:

Students know that rivers and streams can be muddy, especially newly formed ones. But they may not know that some of the clay, sand, and sediment that makes the river muddy comes from the riverbed itself, and is a result of the river eroding or cutting the riverbed. This activity will help students understand the erosion process and its part in the formation of rivers and streams.

Challenge:

Can you build your own model of a river and show how water can cut a riverbed?

Time: 50 minutes

Procedure:

1. Begin a discussion about rivers and streams.

2. Provide each team with an empty half-gallon milk carton. Use scissors to cut out the side panel of the carton under the spout, leaving the spout intact. For younger students, you may prefer to do this cutting for them and then distribute the cartons to the groups. (If water is not available near the outdoor site for the activity, you will have to provide containers to carry water from the classroom sink.)

3. Proceed outdoors. Lay the milk carton on its side, with the cut out panel facing up, and then dig enough soil to fill the container at least half full. Gently pat the soil to smooth the surface.

4. To simulate a flowing river, set one end of the milk carton approximately 1 cm higher than the other end, maybe using a small rock or piece of wood to prop up the carton. The lower end of the carton should be the end with the open spout so when water is poured in at the top end, it will flow over the surface of the soil and out the lower end without forming a "lake."

5. Place the mouth of the bottle on the edge of the higher end of the carton, and slowly pour 2 L of water on the soil, maintaining an even, constant flow of water (see the figure). The goal is to provide a small stream of water, not a sudden flood. Observe what happens to the water and the characteristics of the resulting river, including the path cut and the depth of the riverbed.

If the soil is dry, much of the water will simply be absorbed into the soil, but do not use more than 2 L of water. Runoff

Materials and Equipment:

Each group of two or more students will need:

A 1.9-L milk carton

A 2-L bottle

An outdoor source of soil (ideally a sandy soil)

A graduated cylinder or metric beaker

A metric ruler

Scissors

A water supply

A garden trowel

Q How does water move sediment and soil? How does water shape riverbeds?

Q What was the water carrying down the river? Where did the sediment come from? How did the water cut a riverbed?

of the water in a surface stream will not begin until the soil is soaked with water.

6. Now, repeat steps 3 through 5 with a fresh soil sample, but raise the end of the carton to 3 cm. Be sure to use the same amount of water as in the first trial. Observe the difference in the flow of the water and the resulting river compared to the previous trial.

7. Repeat the procedure for a third time, raising the carton to a height of 5 cm. Compare the flow of water and the river cut to those of the other two trials.

Further Challenges:

You might try building a large stream table by sawing and nailing together wooden boards and using plastic garbage bags as a waterproof lining. A large stream table at a low angle is best for observing a meandering river.

Have students use a variety of soils. Dry sand will quickly absorb relatively large amounts of water, while clay tends to allow the water to run off its surface quickly. If the soil is mixed, students should study what type of soil is most common in the runoff water by pouring this runoff water on sheets of newspaper or paper towels. The soil will remain on the surface of the wet paper.

References:

Stronck, David R. (1971). *Elementary science study. Water Flow.* New York: McGraw-Hill.

Q As the slope of the carton was increased, what happened to the depth of the river formed?

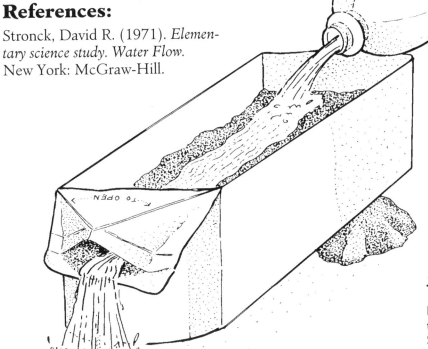

The Author

David R. Stronck is a professor of teacher education at California State University in Hayward and is currently NSTA Research Division Director.

Water, Stones, & Fossil Bones

23 Clean Water: Is It Drinkable?

BY CAROL VANDEWALLE

Focus:

Water is necessary to all living things. Earth is unique in our solar system having about 70 percent of its surface area covered with water. Of this water, only about three percent is potable (fit for human drinking). Of this three percent, almost two-thirds is tied up in glaciers and sea ice leaving approximately one percent of Earth's water available for use by living organisms including humans. This amount of available drinkable water is further reduced by the introduction of pollutants to our water cycle. Clean water is not a limitless resource.

Challenge:

Can you simulate nature's water filtration system and filter out both visible and invisible pollutants from water?

Time: Two to three class periods

Procedure:

1. Cut the plastic bottle in half. Place the mesh over the bottle's mouth and hold it in place with the rubber band.

2. Place the cotton balls, sand, and charcoal into the top half of the plastic bottle, layered as in the figure, making sure to place the cotton over the mesh.

3. Set the top half of the plastic bottle containing the layers of sand and charcoal on top of the lower half of the modified 2-L bottle (see the figure).

4. To prepare the muddy water, fill one of the beakers up to the 100-ml mark with soil. Add water to the 500-ml mark, and stir. Immediately pour it through the filter.

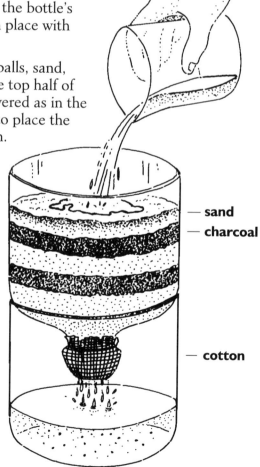

Materials and Equipment:

Each group of two or more students will need:

A 2-L plastic bottle cut in half

A rubber band

An 8-cm square of mesh or nylon netting

500 ml of sand

100 ml of soil

A crushed charcoal briquette

2 cotton balls

250 ml of water

Three 500-ml graduated beakers

50–100 ml of white vinegar

6–9 pieces pH test paper

Food coloring, any color

Chart for recording results

9 small vials or jars such as 35mm film cannisters

✛Safety Note:

Do not drink the filtered water. The nature of this activity may lead students to think the filtered water is fit to drink. Stress to them that it is not. While vinegar wouldn't be toxic in such a small dose, neither would it be very tasty.

Q What do you see? Hear? Smell?

5. Measure the amount of liquid that filters through, and save a small (5–10 ml) sample for reference, labeling it appropriately. Filter this muddy water at least twice more, each time recording how much has flowed through to the bottom half of the plastic bottle, and saving a small (5–10 ml) sample. These small samples will be used to compare turbidity and perform a pH test.

6. Now add 50 to 100 ml of vinegar to the filtered water, test it with the pH test paper, and record the pH level. The pH paper is easier to use than litmus paper and more informative.

7. Now, pour the vinegar-water mix through the filter and into the collection bottle. Test the filtered water again with pH paper and record the data.

8. Repeat the filtering, and test the liquid again.

9. Add several drops of food coloring to the filtered water.

10. Pour the colored water into the filter, and examine the filtered liquid closely. Save a sample of this water. Repeat the filtering twice more with the colored water, each time saving a sample. If you are saving the samples overnight, be sure to cover them to prevent evaporation.

Further Challenges:

Try changing the order of filtering materials, using different combinations of sand, pebbles, crushed limestone, or soils. Ask students which should be on the top? Bottom? Why? Try omitting one of the materials to find out if all are necessary. A unit on acid rain and its effects on plants, stone, metal, lakes, and humans and other animals might effectively follow this lesson.

References:

Calhoun, M. J. (1979). Making water fit to drink. *Science and Children, 16*(7), 48–49.

Gardner, Robert (1982). *The life-sustaining resource: Water.* Englewood Cliffs, NJ: Julian Messner.

Gartrell, Jack E., Jr., Crowder, Jane, and Callister, Jeffrey C. (1989). *Earth: The water planet.* Washington, DC: National Science Teachers Association.

VanDeWalle, Carol. (1988). Water works. *Science and Children, 25*(7), 15–17.

Project Wild—Aquatics. (1987). Boulder, CO: Western Regional Environmental Education Council.

Q Comparing the three samples, what do you think has been happening to the water? Is the water drinkable yet?

Q Do you think the filter can remove unseen substances? Write your predictions down.

Q Did you find any difference in the pH levels?

Q What color will the water be as it filters through and into the collection bottle? Write your predictions down.

Q Comparing this filtered sample with the previous one, what do you observe? What conclusions can you draw about what your filters can effectively remove from the water? What do you suppose water treatment plants do about unseen substances in water such as disease-causing bacteria and harmful chemicals?

The Author

Carol VanDeWalle teaches at the Alwood Elementary School in Alpha, IL.

24 Fred the Fish
An Interdisciplinary Water Pollution Activity

BY PATRICIA CHILTON-STRINGHAM AND JAN WOLANIN

Focus:
Without water, life would be impossible. We use it in many ways—for drinking, bathing, recreation, farming, and manufacturing. We depend on a continuous supply of clean water, yet each time we use it we change it—sometimes by polluting it.

Challenge:
In what ways do we pollute water? How can we clean the water we pollute? How can we prevent water pollution?

Time: One to four 30–45 minute class periods

Procedure:

1. Copy and cut apart the nine roles from the script at the end of this activity, and attach them to the large index cards with tape or glue.

2. Cut a fish shape out of the sponge. Use the yarn needle to thread a string through the bottom of the fish, and then attach the weight so it hangs below the fish.

3. Fill the large glass jar or beaker two-thirds full with cold tap water. Thread another string through the top of the fish, and suspend it in the water by tying it to a pencil positioned across the mouth of the jar. Adjust the length of the string until the fish is suspended midway in the jar of water (see the figure).

4. Number the paper cups or baby food jars 1 through 5, then place soil in cup 1, brown sugar ("fertilizer") in cup 2, pancake syrup ("oil") in cup 3, salt in cup 4, and paper dots ("litter") in cup 5. Pour detergent and warm water into the medium-sized jar, and set out red and green food coloring ("sewage" and "toxic waste").

5. Next, introduce Fred the Fish to the class. Tell them that he has grown up in a protected stream in a nature preserve, but he is about to leave the preserve and journey downstream. The class has been invited to share in his adventure.

6. Distribute the script cards, cups, food coloring, and jar of warm, sudsy water to 17 volunteers. Ask all the students in the class to number their papers from 1 to 9. As the students with script cards read, those with the appropriate ingredients should dump them into Fred's jar on cue. Every student should write down a different descriptive adjective each time

Materials and Equipment:

The whole class will need:

Script pages

A pair of scissors

9 large index cards

A glue stick or some tape

A light-colored sponge

A yarn needle

A small weight (metal nut)

String

A wide-mouthed jar or large beaker

Cold tap water

A pencil

5 small paper cups or baby food jars

Soil

Brown sugar ("fertilizer")

Pancake syrup or molasses ("oil")

Salt

Punched paper dots ("litter")

A medium beaker or glass jar

Detergent

Warm tap water

Red food coloring ("sewage")

Green food coloring ("toxic waste")

they are asked the question, "How is Fred?"

7. After all the ingredients have been dumped in, lift Fred out of the jar, and discuss the change in his appearance and that of the water. (Someone will probably remark that Fred looks dead.)

8. Ask students to compare their lists of adjectives, and then draw cartoons depicting Fred's adventure. (See the example at the end of this activity.)

9. Do not dump the contents of the large jar down the sink. Instead, pour the contents through a strainer over a large, grassy area where natural filtration can take place. Throw away the paper dots strained from the water.

Further Challenges:

Find out where the wastewater in your home or school goes. Contact your local health department regarding septic systems or visit a wastewater treatment plant in your community.

References:

Chilton, Patricia. (1979). *A fish story.* Kalamazoo Soil Conservation District. Kalamazoo, MI.

Q What are some ways to dispose of Fred's polluted water? What are the environmental consequences of each alternative? (Where does water go when it is flushed down the toilet? Poured down the sink?)

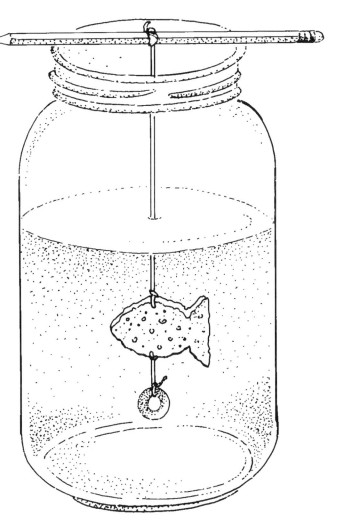

The Authors

Patricia A. Chilton-Stringham is an environmental educator in Portage, MI. Janet L. Wolanin teaches science at the St. Francis School in Goshen, KY.

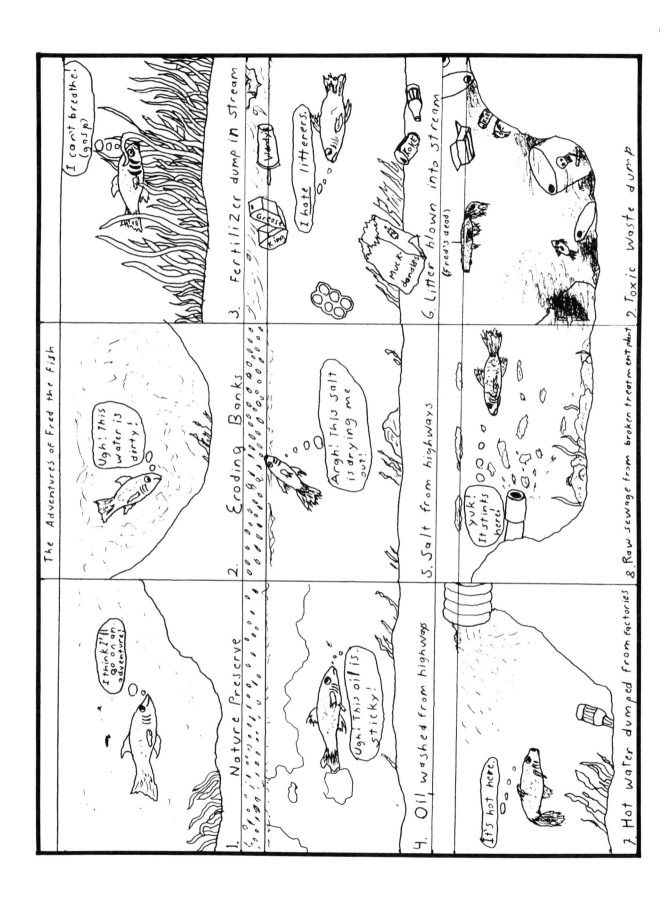

Script Page

1. Imagine a clean river as it meanders through a protected wilderness area. In this river lives Fred the Fish. HOW IS FRED? Fred has lived in this stretch of the river all his life. But now he is going on an adventure and travel downstream.

2. Fred swims into farm country. He passes a freshly plowed riverbank. It begins to rain and some soil erodes into the river. (Dump soil into Fred's jar.) HOW IS FRED?

3. Fred nears a suburban housing development. Some fertilizer from the farms and the lawns washed into the river awhile back. (Place brown sugar in Fred's jar.) The fertilizer made the plants in the river grow very fast and thick. Eventually the river couldn't furnish them with all the nutrients they needed, and so they died and are starting to decay. Their decomposition is using up some of Fred's oxygen. HOW IS FRED?

4. Fred swims under a highway bridge. Some cars traveling across it are leaking oil. The rain is washing the oil into the river below. (Pour pancake syrup into Fred's jar.) HOW IS FRED?

5. During a recent cold spell, ice formed on the bridge. County trucks spread salt on the road to prevent accidents. The rain is now washing salty slush into the river. (Put salt in Fred's jar.) HOW IS FRED?

6. Fred swims past the city park. Some picnickers didn't throw their trash into the garbage can. The wind is blowing it into the river. (Sprinkle paper dots into Fred's jar.) HOW IS FRED?

7. Several factories are located downriver from the city. Although regulations limit the amount of pollution the factories are allowed to dump into the river, the factory owners don't always abide by them. (Pour warm, soapy water into Fred's jar.) HOW IS FRED?

8. The city's wastewater treatment plant is also located along this stretch of the river. The pollution regulations aren't as strict as they should be. Also a section of the plant has broken down. (Squirt two drops of red food coloring into Fred's jar.) HOW IS FRED?

9. Finally, Fred swims past a hazardous waste dump located on the bank next to the river. Rusty barrels of toxic chemicals are leaking. The rain is washing these poisons into the river. (For each leaking barrel, squeeze one drop of green food coloring into Fred's jar.) HOW IS FRED?

25 Soil Leaching

BY GERALD WM. FOSTER

Focus:

Minerals and elements in soil that dissolve in water are removed or leached from soil as water percolates down through it. The minerals or elements remain dissolved in the water unless they react with other substances. If a reaction does occur, a solid may form and no longer be carried by the water; for example, carbon dioxide in water can react with calcium to form deposits of lime. But, if the minerals and elements remain dissolved in the water, they are usually carried into rivers, streams, and lakes.

The amount of leaching depends not only on the composition but also on the particle size of the soil. When leaching and the metabolic activities of plants deplete soil of certain minerals essential for growth—phosphates, potash, and nitrates—fertilizers can be used to replace those minerals. However, excess fertilizers run off the land and contaminate our streams, rivers, and lakes.

Challenge:

Do the pH level of water and the concentration of dissolved minerals in water change after filtration through a soil sample?

Time: 40 to 60 minutes

Procedure:

1. Secure a piece of wire screening over one end of the pipe with the hose clamp, and arrange the PVC pipe and jar as in the figure.

2. Measure the pH of your tap water with the pH paper, and measure the concentration of nitrogen, phosphate, and potash with the soil testing kit. Record these "Before Filtering" measurements (see the sample data table). A pH of 7 is neutral; lower values are acidic and higher values are basic. Use these data as a control to compare with the results from the test samples.

3. Pour a 250 g soil sample into the pipe.

4. Pour 1 L of tap water through the soil sample, and again test the pH and mineral concentration of the water that has filtered through the soil.

5. Empty the soil sample from the tube and rinse it thoroughly. Also rinse out the receiving container. Repeat steps 3 through 4 using the other types of soil, making sure to rinse the tube after each soil sample is tested.

Materials and Equipment:

Each group of four students will need:

50–100 cm of 5 cm (inner diameter) PVC pipe

250 g each of different types of soil such as sand, clay, and humus

A ring stand or other type of support for the PVC pipe

A metric balance

1 L of tap water for each soil sample tested

Wire screening

A three-prong clamp

A hose clamp

A 1-L or larger wide-mouthed container

*pH paper

*A soil testing kit that tests for nitrogen, phosphate, and potash

Paper towels or newspapers

*The pH paper and soil testing kits can be purchased locally at hardware stores and garden supply centers. A county extension service is an excellent source of information on soil and soil conservation. They may be willing to donate soil testing kits for your activity.

Q How do the measurements for the pH, phosphate, nitrogen, and potash contents of the filtered water samples compare for the different soil samples? How do they compare with the tap water before filtering? Have minerals been leached from the soil?

Further Challenges:

Instead of using tap water, repeat the activity with water samples from different sources such as a pond, a stream, or rain.

Another variation is to add various types of fertilizers to distilled water samples, run them through the various types of soil, and compare the pH and the mineral content of the filtered water samples. Discuss what fertilizer runoff can do to nearby streams and water supplies. Caution students to avoid getting fertilizer on their hands or clothes.

You might also ask your county extension service to send a soil scientist to your classroom to give a talk and demonstration.

References:

Schmidt, D. J. (1986). Teaching ecological concepts—cation exchange. *American Biology Teacher, 48,* 406–408.

The Author

Gerald Wm. Foster, Ph.D., is an associate professor of Science Education at DePaul University in Chicago.

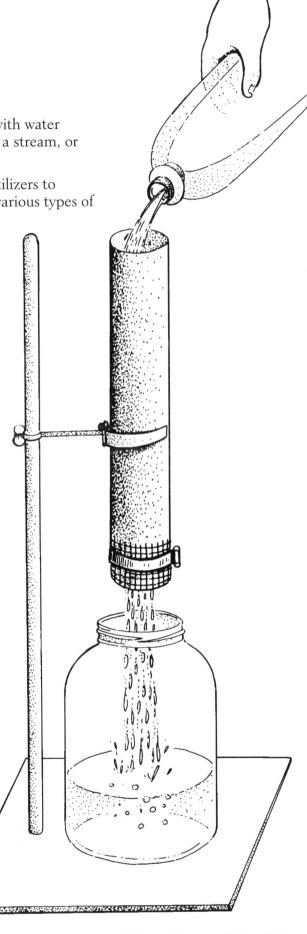

Mineral Concentration and pH Level of Tap Water

Mineral Concentration	Before Filtering Through Soil			After Filtering Through Soil		
	Sand	Clay	Humus	Sand	Clay	Humus
Nitrogen						
Potash						
Phosphate						
pH Level						

When It Rains, It Pours

BY M. ELIZABETH PARTRIDGE

Focus:

Flooding is a natural phenomenon. Its effects can be beneficial, carrying fertile topsoil downstream and depositing it on lands used for farming. But along with the benefits there is often damage to plants animals, property, and sometimes harm to people.

Floods can be caused by overflowing rivers, hurricanes, or tsunamis (sea waves caused by earthquakes). Common causes of river floods are too much rain at one time or sudden melting of ice or snow.

When floods can be predicted, people can protect their houses by using a procedure called sandbagging. Burlap or permeable plastic bags are filled with sand and placed in layers next to the base of the houses. If the flood waters do not rise higher than the sandbags, then the houses will remain dry.

Challenge:

Will sandbagging prevent water from entering our houses? Why or why not? Would we need to sandbag the entire house or just the sides nearest a water source (i.e., a river or canal)?

Time: Two 40–50 minute class periods

Procedure:

1. Give each group a shoe box, scissors, glue, construction paper, and other scrap materials for building a model house. Construction paper strips can be used to cover the outside of the box, forming the exterior, while rectangular pieces may be glued on for doors and windows. The roofs can be constructed by folding a piece of paper in half lengthwise and gluing or taping it to the box lid. Buttons or beads could be glued on as doorknobs and small pieces of green sponge or cellophane added as bushes next to the house. Encourage students to be creative, but make sure they leave the box lid unattached so that they can inspect for flood damage after the experiment.

2. While the glue on the houses dries, fill the burlap squares with sand using the spoons or small scoops. Draw the corners of the fabric square together to form a bundle, and twist tie all four corners of the fabric so that no sand spills out. Ten sandbags per group should be sufficient.

3. When the houses are finished and the glue has dried, move them to the playground near a water source. Arrange

Materials and Equipment:

The whole class will need:

Photos or videos of floods

Three 3.7-L jugs of sand

3 large buckets of water

Each group of students will need:

Ten 25-cm squares of burlap, old sheets, or other permeable fabric

Twist ties

Spoons or small scoops

A shoe box

Construction paper

Scissors

Glue

Scrap materials such as paper, cloth, buttons, etc.

A rock the size of a fist

the houses into a neighborhood at the bottom of a slight incline, and place a rock in each house so it won't be carried downstream by the flood. If no incline is available, construct one by propping a piece of plywood on a cement block or create one with potting soil or dirt.

4. Fully sandbag one-third of the houses by tightly packing bags all around the houses in two layers. Another one-third of the houses should be partially sandbagged along the sides nearest the rising flood waters, and the remaining houses should be left unprotected (see the picture).

Q Can you predict how much damage our flood will cause each type of sandbagged house?

Q How did the sandbags protect the houses?

5. To simulate a flood, pour two or three large buckets of water steadily down the incline toward the model neighborhood.

6. Following the flood, remove the roofs and inspect the houses for interior damage. The fully sandbagged houses should be dry, the partially protected houses should show some water damage, while the unprotected houses should be flooded.

Q What do you notice about the sand? Why did the sandbags keep the fully sandbagged houses dry? If you had to sandbag a house, would you sandbag around the entire house or just on the sides where the water is coming from? Why?

7. Next, inspect the sandbags to see why they protected the house.

Further Challenges:

Repeat the activity using plastic sandwich bags for the sandbags. Why do plastic bags make better or worse sandbags than permeable fabric in a real flood? How could houses be built to protect them from floods? Students could report on famous floods such as Johnstown, Pennsylvania, or Hurricane Camille, or investigate the positive aspects of flooding, such as improving soil.

The Author

M. Elizabeth Partridge is an associate professor at Southeastern Louisiana University in Hammond.

Grinding and Scraping

BY KAREN K. LIND

Focus:
A glacier is a large accumulation of flowing ice which often incorporates rocks, plants, and soil from its path into its structure. Pressure causes the rocks and sediment at the bottom of the ice to scour the land, leaving scratches in rocks beneath the glacier.

Challenge:
Do you think that water can scratch surfaces? Make an ice cube glacier and find out.

Time: 30 minutes to prepare the ice cube trays, several hours to freeze the cubes, and another 45 minutes to "grind and scrape"

Procedure:
1. Pour water into the ice cube trays and let students add varying amounts of sand, twigs, and pebbles to half of the cubes. Keep the other half of the ice cubes free of debris for comparison.

2. After the cubes freeze, give each student one clear cube and one with debris. Rub the clear ice cube in one direction over a flat piece of aluminum foil and then do the same with the debris-filled cube on another piece of foil and compare.

3. Repeat step 2, but use wood scraps instead of foil for the surface. Does the ice scratch the surface or remove any material from it? How hard do you have to press the ice to make an impression on each of the materials?

4. Let the ice cubes melt on a paper towel without disturbing them, and consider how the remaining sand and pebbles compare to the boulders and sediment left by glaciers.

Further Challenge:
Try scratching the two different types of ice cubes over modeling clay rolled out flat and the backs of ceramic tiles, and then examining these materials after being scraped. Does the ice scratch the surface or remove any material from it? How hard do you have to press the ice to make an impression on each of these materials? Show pictures of glacial deposits, and ask your students to describe how large the glacier must have been that formed them. Consider how heavy glaciers must be to carve out impressions large enough to make valleys, rivers, and lakes.

Materials and Equipment:

The whole class will need:
Several ice cube trays
Water
Freezer
Twigs, pebbles, and sand
Aluminum foil
Modeling clay
Wood scraps
Ceramic tiles
Paper towels

Q What happened to the foil with the first and second cubes? What difference would more sand make in the ice cube?

Water, Stones, & Fossil Bones

The Author

Karen K. Lind is an assistant professor in the school of education at the University of Louisville in Louisville, KY.

References:

Gartrell, Jack E., Jr., Crowder, Jane, and Callister, Jeffrey C. (1989). *Earth: The water planet.* Washington, DC: National Science Teachers Association.

Lind, Karen (1989). A lesson on rocks and weathering. *Science and Children, 26*(8), 32–33.

McBiles, J. L. (1985). *My home....the Earth.* Nashua, NH: Delta Education, Inc.

Drip Sculpture

BY EDWARD P. ORTLEB

Focus:
Underground water seeping through porous rock and cracks carries dissolved minerals. As the water drips from the ceiling of a cave, some of it evaporates and leaves behind a mineral deposit in the form of a stalactite. Water that drips onto the floor of a cave also evaporates, and this mineral deposit forms a stalagmite. But stalactites and stalagmites grow very slowly: large cave formations take thousands of years to form. Because of this, young students sometimes have difficulty visualizing how stalactites and stalagmites form. This activity allows them to witness model formations over the course of just a few days.

Challenge:
How do dissolved minerals form cave deposits like stalactites and stalagmites? Can you create stalactite- and stalagmite-type formations in a much shorter time?

Time: Three to five days

Procedure:
1. Into a large container, pour an amount of hot water equal to the volume of the two small glass containers.

2. While stirring continuously, add Epsom salt to the water until no more salt will dissolve. Do not be surprised by how much it will take.

3. Fill each of the smaller containers with the concentrated Epsom salt solution to within 2 cm of the brim.

4. Place the two small containers in an area where they will not be disturbed and position them about 8–10 cm apart.

Materials and Equipment:

Each group of two or three students will need:

Epsom salt

Hot tap water

A large container for mixing

A spoon

2 small glass containers such as beakers or baby food jars

A piece of heavy cotton cord or string, 15–20 cm long

Aluminum foil

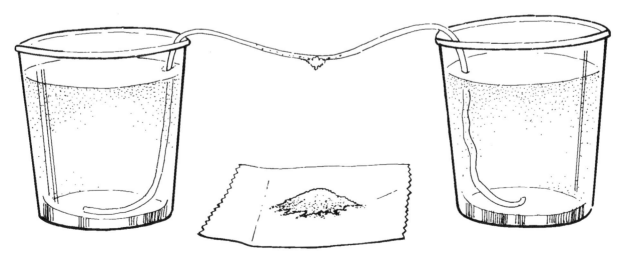

Water, Stones, & Fossil Bones

Q What are stalactites and stalagmites made of? How are columns formed in caves? How might several years of drought affect the rate of stalactite formation?

The Author

Edward P. Ortleb is Science Supervisor for the St. Louis Public Schools.

Insert the ends of the string into the containers and let the string sag in the middle (see the figure). You may have to put paper clips on the ends of the string to keep them underwater. The Epsom salt solution will move along the cord from both sides and form a dripping spot at the sagging center of the cord. Place a piece of aluminum foil under this area.

5. Let the containers stand undisturbed for several days. As the salt solution evaporates, the Epsom salt will begin to deposit on the string in much the same way a stalactite forms. Within three days, a small stalactite will form. If left undisturbed for a week, it will be approximately 1 cm long. A small stalagmite may also form on the aluminum foil below.

6. Explain to your students that natural cave formations take much longer because the amount of dissolved mineral material (calcium carbonate) in the dripping water is very small and evaporation is usually very slow.

Further Challenges:

Use other kinds of substances to make the formations, such as table salt or sugar. Will the results be the same?

References:

Navarra, J. G. and Zafforoni, J. (1963). *Science today for the elementary school teacher.* Evanston, IL: Harper and Row.

Earth Science—Asking Questions and Giving Instructions

Teachers ask many types of questions and direct students along different avenues to find answers. Teachers' questions may be divergent or convergent. Divergent questions and instructions do not have one right answer or response. They allow for a variety of answers and lead students to think along many different lines. For example, questions that begin with "Tell me about . . .," "What do you think . . .," " What have you found out . . .," "What can we do with . . .;" and instructions such as "You can examine these . . .," or "You may explore these . . ." provide an opportunity for being creative, predicting, and experimenting.

Convergent questions or instructions, on the other hand, ask for a specific response or activity such as "How many raindrops can you count?" "Name the type of clouds," or "Find a rock smaller than this one." Teachers often ask only convergent questions and give convergent instructions that permit fewer opportunities for creativity and investigation. Questions and instructions should invite divergent responses as much as possible because students need time to construct their own ideas. Activities such as "Catch a Raindrop" and "Hot Days, Cold Days" engage both types of questions and instructions and recommend strategies that encourage students to think and act for themselves.

Often how a teacher frames a questions will determine whether the student thinks divergently or convergently. Note the difference between "What did we study yesterday?" and "What did you find out yesterday?" Both questions require a review of a previous activity, but the former seeks an answer already in the teacher's mind while the latter probes the student's own experience.

29 Hot Days, Cold Days
Graphing Daily Temperature

BY MAUREEN J. AWBREY

Focus:
The following activities enable kindergartners to observe changes in daily temperature through the use of a bar graph.

Challenge:
How much does the temperature vary from day to day within the space of a month? How can we record it so the changes can be easily seen?

Time: 10 minutes a day for a month

Procedure:
1. Choose a month in which you can expect some major fluctuations in the daily temperatures, such as January, February, or March. If the thermometer has two scales on it, one for Fahrenheit and another for Celsius, cover the Fahrenheit scale with opaque tape.

2. Put a horizontal red line on the thermometer just above the bulb.

3. Place the thermometer outside the window in a consistently shady place, either by setting it on the window ledge or by suspending it with string. Put the bar graph on a bulletin board or on the wall near the calendar, and place the paper strips, scissors, glue, and chalk in a box nearby.

4. The temperature should be recorded at about the same time every day. When it is time to record the daily temperature, ask one student to take a red strip and the chalk from the supply box. Bring in the thermometer, quickly place the strip with its bottom edge on the red line, and make a mark on the strip beside the top of the alcohol column. Place the thermometer back outside.

5. Cut the red strip at the mark, and glue it to the bar graph, placing it in the correct column for the day of the month. The height of the strips is a graphic indication of temperature changes from one day to the next.

Further Challenges:
You may want to write the actual temperature at the bottom of each day's strip. Reading the numerals on the thermometer requires help from the teacher, as it involves the skill of counting by two.

As an added dimension to the activity, mark the graph at the

Materials and Equipment:
The whole class will need:

An oversized alcohol thermometer that is easy to read

A bar graph with columns for each day of the month

Strips of red construction paper to fit the bars on the graph

Chalk

Scissors

String

A small bottle of glue

A marker

✢Safety Note:
If it is below freezing outside and the thermometer is metal, be sure the student bringing in the thermometer has dry hands, or better yet, is wearing gloves.

freezing point with a red horizontal line. Then, as each student glues on the strip, he or she can discuss whether it is above or below freezing on that particular day. At the end of the month, the students can count how many days the temperature was below freezing.

Ask the students why the graph from the afternoon class often has longer bars than the morning class's graph?

References:

Baratta-Lorton, M. (1976). *Mathematics their way*. Menlo Park, CA: Addison Wesley.

Q Is it growing warmer or colder as the month goes on? How does the temperature affect how we dress? In what ways does the temperature affect our daily activities?

The Author

Maureen Jessen Awbrey is a kindergarten teacher at the Anchorage Independent School in Anchorage, KY.

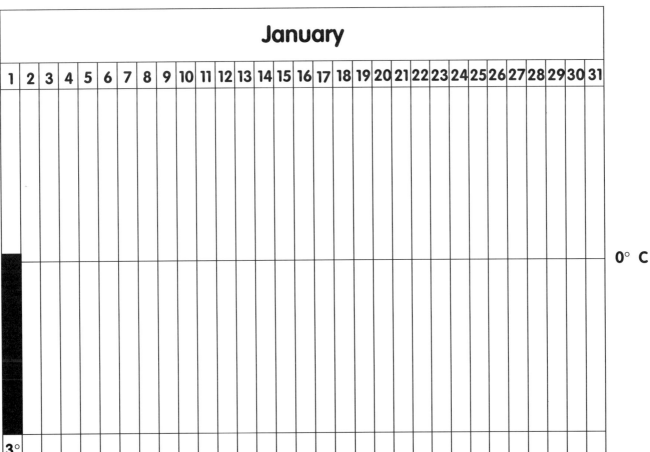

30 Capturing Heat from the Sun

BY VINCENT G. SINDT

Focus:

How many times have you been surprised by your students when you assumed that they understood simple concepts and you found they did not? These surprises often happen with concepts concerning solar energy. Building a simplified solar collector and performing the following series of activities will help students understand how to better capture heat from solar energy.

Two major components of the collector constructed in this activity are the absorber plate and cover plate. The absorber plate absorbs sunlight and re-radiates its energy as heat. The cover plate serves two purposes: first, it allows sunlight to pass through and strike the baffle, and second, it traps the heat radiated from the absorber plate inside the collector.

Materials and Equipment:

Design may vary. The whole class will need for two collectors:

A sunny day

Cardboard or masonite

Glue, screws, or nails

A variety of colored, transparent plastic sheets 29 cm x 41 cm

A clear plastic sheet 29 cm x 41 cm

2 transparent fiberglass sheets (patio fiberglass) 0.3 cm x 29 cm x 41 cm

A rippled fiberglass sheet (greenhouse covering) 29 cm x 41 cm

Aluminum foil

Polystyrene insulation board 1.9 cm x 26 cm x 37 cm

4 carriage bolts with washers and wingnuts

Small cans or baby food jar lids

Black construction paper

White construction paper

Flat black spray paint

White spray paint

A protractor

3 metal thermometers (one for outside temperature, two for inside the collectors) (range 0–110°C)

A temperature log book

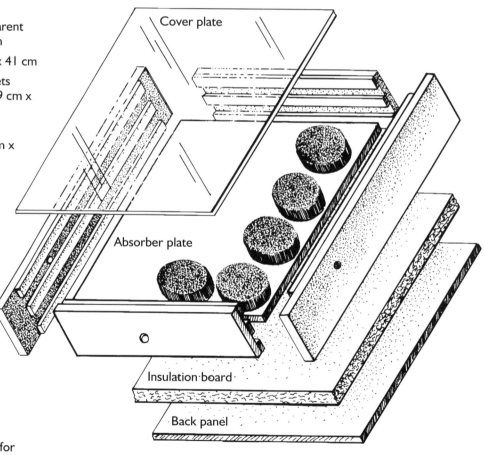

figure 1

Several characteristics of solar collectors help explain how they gain and transfer heat. Some of these characteristics include the following:

70 National Science Teachers Association

- Dark colors absorb more sunlight than light colors, therefore dark colored baffles radiate more heat than light colored ones.
- Increasing the surface area of the absorber plate will increase the amount of sunlight it can absorb and therefore the amount of heat it can radiate.
- An enclosed and insulated collector traps the heat radiated from the absorber plate so it can be harnessed.
- The collector's orientation to the Sun affects the amount of sunlight absorbed.

Challenge:

Can you build an efficient solar collector? How will the following design features of the collector affect the temperature inside of the collector: the type of cover plate, the texture and color of the baffle, and the collector's orientation to the Sun?

Time: One to two classes to build the collector and 2–1/2 hours for each experiment

Procedure:

1. Depending on the ages of your students, you may construct the collectors yourself or involve the students in the process.

It is best if two collectors are used for these experiments. Students will need to compare the interior temperatures of a collector with different absorber and cover plates. If only one collector is used, by the time the plates are changed a cloud could appear and obstruct the sunlight, and the comparison of the temperatures would therefore be invalid. With two collectors, the temperatures can be recorded at the same time.

With older students, ask for suggestions on how to alter the design of the collectors to make them more efficient. Important features of this activity's collector design are that the cover and absorber plates are replaceable and the collector can be set at different angles. Encourage students to come up with different absorber plates and cover plate materials. Once the collectors are complete, try several or all of the experiments outlined below.

Conduct the following experiments when the outside temperature is about 25°C. Unless it is otherwise specified, use a

Water, Stones, & Fossil Bones

clear cover plate and a smooth, black absorber plate.

Experiment 1: Color of the absorber plate.

1. Staple or tape black construction paper to cover one absorber plate and white construction paper to cover the other plate.

2. Set the collectors side-by-side facing the Sun, and allow the collectors to warm up for about 30 minutes. Then record the temperatures from both collectors every 30 minutes for 2 hours and calculate an average temperature for each collector.

Q Why is the collector with the black absorber plate hotter than the one with a white absorber plate?

3. Compare the temperatures of the two collectors.

Experiment 2: Surface area of the absorber plate.

1. Cover both baffles with aluminum foil, and spray paint them with flat black paint. Also paint enough short cans, such as tuna, or baby food jar lids, to cover a absorber plate. (Steel wool painted black could be used instead of the cans.)

2. Glue or screw black cans to one of the baffles. (Watch for sharp edges on the cans.)

3. Place the collectors side-by-side facing the Sun, and allow them to warm up for 30 minutes. Then record the temperatures from both collectors every 30 minutes for 2 hours and calculate an average temperature for each collector.

Q How does increasing the surface area affect the temperature inside the collector? Why?

4. Compare the temperatures of the two collectors.

Experiment 3: Color of the cover plate.

1. Overlay one cover plate with the clear plastic sheet, and the other cover plate with one of the colored plastic sheets.

2. Place the collectors side-by-side facing the Sun, and allow them to warm up for 30 minutes. Then record the temperatures from both collectors every 30 minutes for 2 hours and calculate an average temperature for each collector.

Q What effect does the colored plastic covering have on the temperature inside the solar collector? Why?

3. Compare the temperatures of the two collectors.

4. Repeat this experiment with other colors of plastic.

Experiment 4: The texture of the cover plate.

1. For one of the cover plates, use rippled fiberglass or greenhouse covering.

2. As in the other experiments, place the collectors side-by-side facing the Sun, and allow them to warm up for 30 minutes. Then record the temperatures from both collectors every 30 minutes for 2 hours and calculate an average tem-

perature for each collector.

3. Compare the average temperatures of the collectors.

Experiment 5: The angle toward the Sun.

1. Place the same type cover and absorber plates in the collectors.

2. Stand one collector vertically and the other at a 10° angle from vertical, both oriented toward the Sun. Allow them to warm up, and then record the temperatures from both collectors every 30 minutes for 2 hours.

3. Repeat this experiment three times placing one collector at angles of 20°, 30°, and finally 40° while leaving the other collector vertical.

4. Compare the average temperatures of the collectors for each run of the experiment.

Q Why is the collector with the smooth cover plate hotter than the one with the rippled surface?

Q Why is the angled collector hotter than the vertical one?

Further Challenges:

Remove the cover plate of one of the collectors and measure the effect of trapping versus not trapping the heat radiated from the baffle. Or place blocks inside of one collector to change its volume. As another variation, place a container of water inside of one of the collectors to see if vapor pressure or moisture has an effect on the temperature. Finally, partially cover the front of one of the collectors with paper, and study how the size of the area through which light enters affects the temperature of the collector.

The Author
Vincent Sindt is an associate professor of natural science and science education and Director of the Wyoming Institute for the Development of Teaching at the University of Wyoming in Laramie.

31 Cooking with the Sun

BY RUTH M. RUUD

Focus:

Future generations will have to discover and use new sources of energy to meet their needs. Students should become aware of the shortages of natural resources and the need to explore alternate sources of energy. One alternate source is solar energy. This energy can be transformed into electricity by solar cells such as those on a solar calculator. Infrared light from the Sun provides yet another form of energy—heat. Ask your students how many have used a solar clothes dryer (drying clothes outside on a line).

The amount of solar energy a location receives is determined by the Sun's elevation (dependent on the latitude and season); the length of the day (dependent on the season); and the amount and number of days of cloud cover. Because of the variability of these three factors, we need to use technology to amplify the energy from the Sun.

In this activity students will design and construct a solar cooker which will amplify the heat from the Sun.

Challenge:

Can the Sun be used to heat and cook hot dogs? Try building a solar oven and roasting a hot dog in it. How would you change the design of this oven to make it more efficient? What variables would have to be considered in order to use solar power in your climate? What materials would be best to use?

Time: Two hours

Procedure:

1. Working from the diagram provided, completely cover the inside of the box with aluminum foil. Poke the skewer through one side of the box, put the hot dog on, then poke it through the other side. The skewer can be raised and lowered by making a new pair of holes in the sides of the box. You may want to use tape to hold the skewer in place.

2. Go outside and select a site. Students should tilt their cooker toward the sun by propping up one end of the cooker with a rock or other object. Record the time of day and the direction the cooker is facing on a record sheet. Students may need to adjust the angle depending on the Sun's elevation.

3. Check the hot dogs every 20 minutes and record observations. After one hour check to see which hot dogs are

Materials and Equipment:

Each group of two or three students will need:

A diagram of an oven (see the figure)

A shoe box

Aluminum foil

A twig or stainless steel rod

Scissors

Tape

A hot dog

A watch

A compass

A record sheet

A sunny day

✚Safety Note:

Be sure to cover the front of the box with plastic wrap if the hot dogs are to be eaten. Also use clean material and a stainless steel skewer that has been washed.

cooked, raw, or burned. The time will vary with the climate and latitude. Ask why some hot dogs cooked faster than others. Have students draw conclusions based on their observations. You may want to award certificates for best cook and best cooker.

4. Now, have each group analyze their results and decide which qualities of their cooker were advantageous and which were disadvantageous. In their evaluation, they should consider the design, site, and direction their cooker faced.

5. Present the analyses and, as a class, use that information to design the best possible cooker.

Further Challenges:

Modify the cookers even further by adding mirrors or reflecting panels, or try more panels at wider angles. Then, have the students try cooking an egg in their new and improved solar cookers. Have students study the Sun's elevation, the seasons, and the latitude to determine if there would be a better time of year and day to cook the hot dog.

The Author

Ruth M. Ruud is a teacher at the Chestnut Hill School in Erie, PA.

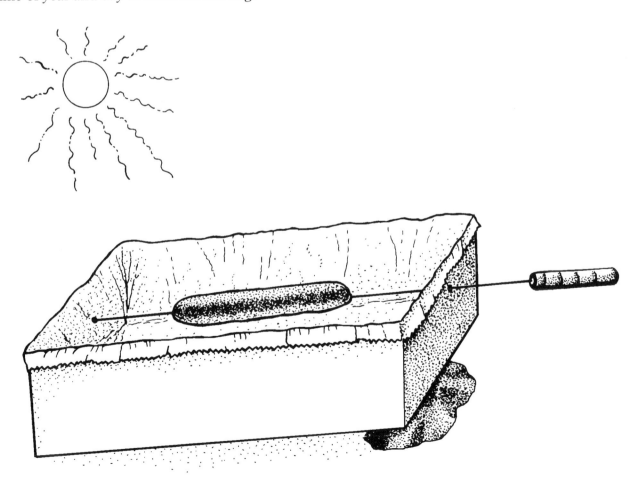

Water, Stones, & Fossil Bones

32 Catch a Raindrop

BY GERALD WM. FOSTER

Focus:
Not all raindrops are the same size and students can prove it by capturing and examining individual raindrops.

Challenge:
Are all raindrops the same size? How could you measure a raindrop? Can you catch a raindrop without destroying its shape?

Time: 40 minutes

Procedure:

1. Wrap plastic wrap around the outside of the strainer and secure with a rubber band or tape. The plastic wrap prevents the flour from falling out as students carry their strainers outside and then back to the classroom.

2. Put a layer of flour approximately 1 cm thick in each strainer. There should be enough flour to soften the impact of the raindrops and to prevent penetration to the wire mesh. Window screening will work just as well as a strainer. A large strainer or piece of window screen provides an area for raindrops to hit without running together right away. You may want to try different sized meshes to see which one allows the flour to sift through without breaking raindrops.

3. Students may work in groups, one being a timer, another counting the drops as they hit, and another responsible for placing the collector at a particular height. Several trials may have to be made to obtain consistency in data and/or an average of the class's data.

4. Go outside, and allow raindrops to land in the strainer. Stay out just long enough to catch a few good samples; staying out too long will only create a pasty mess.

5. Carry the strainer back inside, remove the plastic wrap, and gently shake away the loose flour. The raindrops will be encased in flour, retaining their shape.

Materials and Equipment:
Each group of two or three students will need:

A rainy day

Flour

A large kitchen strainer (with flat bottom)

A metric ruler

Plastic wrap

A rubber band or tape

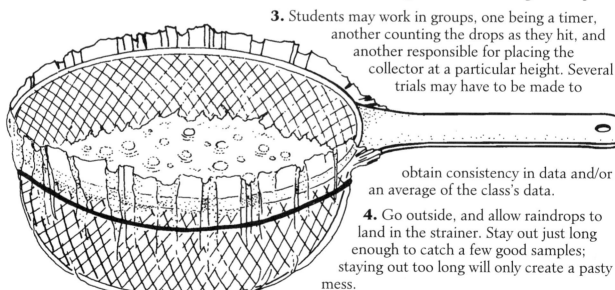

6. Next, compare the sizes of raindrops. Older students can measure their drops with a metric ruler, but all students can develop a chart for the following data: number of raindrops that can be caught before they run together in the flour, number of different sizes of raindrops, number of each size, and for older students, the actual sizes of the raindrops. In order to have meaningful data, consider all of the variables that could affect the results; for example, surface area of the raindrop collector, size and shape of collector, where the collector is placed, and how long the collector is left in the rain. These are variables that students can control.

Further Challenges:

Can the students make different sized water drops and catch them in the flour? Are there other ways to catch raindrops besides using flour? (Try colored construction paper.) What kinds of patterns are made by raindrops or water drops when they hit pieces of colored construction paper? Devise a way to prove whether or not one gets more wet walking in the rain or running through it. Put your raindrops in the freezer. When they are frozen, weigh them and compare their shapes and sizes.

References:

Schaefer, Vincent J. and Day, John A. (1981). *A field guide to the atmosphere.* Boston: Houghton-Mifflin.

The Author

Gerald Wm. Foster, Ph.D., is an associate professor of Science Education at DePaul University in Chicago.

33 Fakey Fog?

BY GERALD WM. FOSTER

Focus:

Fog is the result of water vapor in the air condensing on the surfaces of air particles. For this to occur, the air must first become saturated, or have a relative humidity of 100 percent. This can happen when warm, moist air is cooled by passing over land, water, or another air mass that is cooler than it is. Then, if the saturated air is cooled any further below this saturation temperature or dew point, the water molecules condense, combining with each other and collecting on a surface. For condensation to occur, there must be a surface available. In the formation of fog, this surface is the array of particles suspended in the air.

Challenge:

How does fog form? What conditions are needed to produce fog? Can you produce fog in a jar?

Time: 30 minutes

Procedure:

1. Tell the students they are going to simulate the production of fog, and encourage them to tell their partners their observations.

2. Put a spoonful of water in the jar, attach the lid, shake the jar, and empty the water out of the jar. This raises the humidity and temperature in the jar, aiding in fog formation.

3. Light a match and drop it into the jar. As soon as the match goes out, cover the mouth of the jar with the lid and place an ice cube on top of the lid (see the figure).

4. The air inside the jar will cool and sink toward the bottom, so students should observe fog starting to form as the air cools below its saturation temperature. The water vapor will be condensing on the suspended smoke particles. Fog will form at the top of the glass container and slowly fall toward the bottom of the container.

Further Challenges:

What weather conditions are needed to form fog? Are there certain times of the year when fog is most likely to occur? Does the shape of the land influence fog formation? Why would you expect to find fog occurring in mountainous and hilly areas compared to flat land?

Have the students observe what happens when they place their jars of fog in hot water. Ask what they think will happen

Materials and Equipment

Each group of two students will need:

A small, narrow-mouthed jar with a lid

Water

Matches

Ice cubes

Safety goggles

✣ Safety Note:

Be cautious when handling glass and lighted matches. Wear safety goggles when striking matches, and provide containers (jar lids will do) for burned matches.

Q How does the formation of fog compare with that of clouds?

to the fog if they put their jars in a freezer, and encourage them to try this at home with parental supervision. Can they develop another way to produce fog?

References:
Victor, Edward. (1989). *Science for the elementary school.* New York: Macmillan.

The Author

Gerald Wm. Foster, Ph.D., is an associate professor of Science Education at DePaul University in Chicago.

34 A Simple Hygrometer Mobile

BY DENISE TASSI-KANE

Focus:
Cobalt Chloride is a chemical which indicates the presence of water by changing color. When there is water vapor in the air, cobalt chloride turns pink, but it turns blue when the air is dry. A device that detects humidity is called a hygrometer.

Challenge:
Using a simple hygrometer, can you predict the weather?

Time:
30 minutes to construct the mobile and then five minutes a day for 14 days to record data

Procedure:
1. Use the stencils provided or design your own mobile, and cut them out from cobalt chloride paper.

2. Make a hole in each of the shapes, and connect the stencils together with string to make a mobile.

3. Hang the mobile and then, at the end of each day for 14 days, record the color of the mobile and what the weather is like on that day. Record this data at least until the weather changes.

Materials and Equipment:
The whole class will need:
Paper treated with Cobalt Chloride
Scissors
String
A raindrop template
A cloud template

Q Can you predict, from the color of the mobile, what the weather will be like on a particular day?

Further Challenges:

You may want to examine more systematically the relationship between the water vapor content in the air and the color change of the cobalt chloride paper. Simply place pieces of the coated paper in different environments such as a closed jar with silica gel, a jar with an open container of water in it, in front of a humidifier, in a frost free refrigerator, etc.

References:

Franklin Institute. (1986). *Make a simple hygrometer mobile*. Philadelphia, PA: Author.

The Author

Denise Tassi-Kane is Coordinator of Teacher Resources for the Museum-To-Go! Resource Center at the Franklin Institute in Philadelphia.

35 Moving Air
The Whoosh Box

BY JUDY PESSOLANO

Focus:
Even though we cannot see, hear, smell, or taste air, it is all around us. Air takes up space, and we can feel it when it moves or we move through it.

Challenge:
How do we know that air exists when we cannot see, hear, smell, or taste it? Can you find a way to make air move so you can see it?

Time: 30 minutes

Procedure:

1. As a demonstration for the class, cut a hole about 2 cm wide and 10 cm long in the top of the empty cereal box. Rather than trying to punch a hole in the top, simply mark the top where the hole is to be, open the box, and cut into the edges of the box tops so that when you fold the ends back down, you have cut the right size hole. Make sure to tape the flaps down so the top can't fly open

2. Wrap the box with tissue paper, but do not cover the hole.

3. Cut streamers about 1 cm wide and 5 cm long, and tape the tissue streamers above the hole so that they hang across the hole.

Materials and Equipment:

The whole class will need:

An empty cereal box

Tissue paper—enough to cover the box and make streamers

Cellophane tape

A metric ruler

Scissors

A pen or pencil

✜Safety Note:

If your students are each making their own "Whoosh Boxes," be sure they do not try to punch a hole in the box, but rather cut into the edges of the box tops as explained.

4. Pass the box around, and let the students shake and feel the box. Ask them "What is inside of the box?" Have the students squeeze the box in the center. They will see the tissue streamers move about over the hole.

The students should conclude that air was in the box, and when it was squeezed some of the air came out of the box causing the tissue streamers to flutter around.

Further Challenges:

Bring in several bottles of play bubbles. Have the students take turns blowing bubbles. Ask "What is inside of the bubble?" "Where did the air inside of the bubble come from?"

References:

Eliason, Claudia F., and Jenkins, Loa T. (1981). *Early childhood curriculum.* St. Louis: The C.V. Mosby Co.

Q If nothing is inside the box, what made the tissue paper move?

The Author
Judy Pessolano is a kindergarten teacher at the Kentucky Country Day School in Louisville.

36 When Is It Dew?

BY SHIRLEY G. KEY

Focus:

Condensation is the changing of a gas (vapor) to a liquid. This change can be triggered by a drop in temperature; for example, when water vapor in the air condenses as dew. Warm air holds more water vapor than cool air, so that as the air cools, it becomes saturated and the relative humidity reaches 100 percent. As air cools beyond the saturation point, water vapor begins to condense on surfaces. The temperature at which the air is saturated is known as the dew point and the water that condenses below that is called dew.

Challenge:

What is condensation? What causes a mirror to fog up when you breathe on it? What is happening when you see your breath on a cold winter day? How and when does dew form? Can you find the dew point for condensation that forms on a can of cold water?

Time: 45 to 55 minutes

Procedure:

1. Introduce this activity by discussing weather, the water cycle, and condensation.

2. As a preliminary activity, have the students blow into the air and ask them what they see (usually nothing). Have them breathe on a piece of aluminum foil or a mirror. They should see some mist on the mirror or foil. Now have the students blow up balloons and ask them what is happening and what is collected in the balloon.

If students understand that air can be collected, it will be easier for them to grasp the concept that air holds water vapor that also can be collected.

Students should continue to observe the water droplets on the foil or mirror. Discuss how this moisture got on the surface of these items.

3. For the main activity, open a discussion with these questions: Where does the water from the ocean go when it evaporates? Where does the water from a bathtub or shower go when it evaporates? Where does the water on plants, cars, and houses that we see in the early morning come from? What type of temperature change must occur before condensation or dew forms? And finally, ask them what they think the dew point is.

4. Now, fill the can half full with water. Add four to six ice

Materials and Equipment:

Each group of three students will need:

A metal can

A thermometer

Ice cubes

A stirring rod

Water at room temperature

A beaker

3 balloons

A piece of aluminum foil

Q Why can't you see moisture on the foil until you breathe on the surface? Under what conditions will moisture collect on those surfaces?

cubes to the water, stir, and put the thermometer into the water.

5. To see the condensation when it is first forming, continually wipe the side of the can with your finger until you feel moisture or you see a path left by your finger through the condensation forming on the outside of the can. When you first notice the condensation, quickly measure the temperature and write it down. This temperature is very close to the dew point.

6. All groups in one room will usually get temperature readings very close to each other. After seeing condensation form on the cans, students should be able to visualize how dew forms in our environment. Be sure they do not form the impression that the vapors penetrated to the outside of the can. Emphasize that condensation on the can came from water vapor in the air around the can.

7. Have each group read their dew points to the class. Explain that both the temperatures of the surrounding air (which vary around the room) and the temperature of the ice water affect the dew point.

Further Challenges:

Ask students what they think happens to dew when the temperature continues to drop. From their observations of this activity, can they guess how rain forms?

References:

Bernstein, L., Schacter, M., Winkler, A., and Wolfe, S. (1986). *Concepts and challenges in Earth science.* (2nd ed.) Englewood Cliffs, NJ: Globe Book Company.

The Author

Sheila Gholston Key was a science teacher at Christa McAuliffe Middle School in Houston, TX and is currently a doctoral student at the University of Houston.

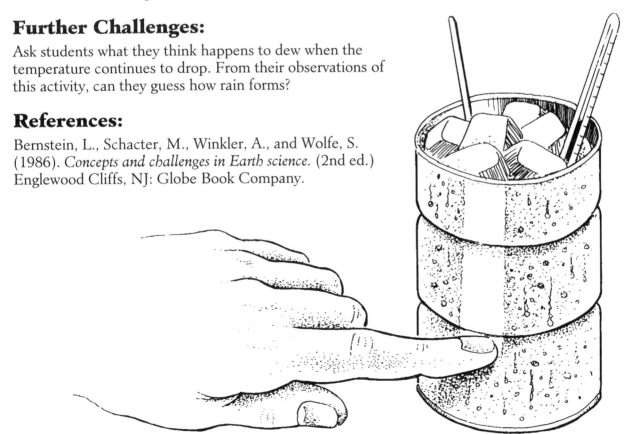

Water, Stones, & Fossil Bones

37 Dew or Frost?

BY MERRICK OWEN

Focus:
Cooling air can cause several forms of condensation including clouds, fog, and dew. Dew is formed when warm air carrying water vapor touches a cold surface and the surface is above 0°C. If the cold surface is at or below 0°C, frost will form. This activity is best done in the spring or fall when the relative humidity is 50 percent or more.

Challenge:
What will form on the outside of a can when you add ice cubes to the water? What will form when you add salt to the ice cubes?

Time: 50 minutes

Procedure:
1. Fill the can one-third full of water that is about room temperature. Place the thermometer in the water and record the temperature.

2. Add crushed ice until the can is half full and record the water's temperature every 30 seconds until condensation appears on the outside of the can. Be sure not to hold the can by the sides.

3. For the next part of the activity, empty the can and dry the outside. Place the thermometer in the can, fill one-third of the can with crushed ice, and record the temperature of the ice.

4. Cover the ice with salt, record the temperature, stir the salt into the ice, and record it again every 15 seconds until 2 minutes have passed. During this time, observe the moisture on the outside of the can.

Explain that adding the salt made the ice melt faster, causing it to absorb more heat and lower the surrounding temperature to below 0°C, the freezing point of water, thus forming frost on the can.

Further Challenges:
Calculate and graph the average temperature from several trials of the above activities.

Materials and Equipment:

Each group of two or three students will need:

A metal can

A thermometer

Cracked or crushed ice

Salt

Water

A paper towel

A data sheet

Q At what temperature did condensation form on the outside of the can? What do you call this moisture when it forms on grass?

Q What happened to the moisture on the can? What do you call this moisture when it forms on grass?

The Author

Merrick A. Owen, Ph.D., is Planetarium Director and a professor of geosciences at Edinboro University of Pennsylvania in Edinboro.

Something in the Air

BY M. ELIZABETH PARTRIDGE

Focus:

Air pollutants are chemicals (in gaseous or liquid form) or suspended particles that have reached concentrations harmful to living organisms and/or materials such as stone and metal. Most air pollution results from human activities such as burning fuels for transportation and industrial purposes, and disposing of waste. But there are natural sources of air pollutants as well such as pollen, dust, and volcanic ash.

Dust and volcanic ash may be visible as air pollution when they are in high concentrations. But a more common form of air pollution that is highly visible is smog—a combination of smoke and fog or the result of an interaction of certain pollutants with sunlight.

But regardless of whether air pollution is visible or not, it can cause respiratory problems in people and other animals, leaf damage or death of vegetation, accelerated deterioration of statues and buildings, and depletion of the ozone layer in the atmosphere.

Challenge:

What is air pollution? Can we always see pollutants in the air? How can you capture air pollution particles? Does the amount of air pollution differ at various locations around your school? Why is the air in some areas, like large industrial cities, more polluted than others?

Time: Two 25-minute sessions

Procedure:

1. Show pictures of severe air pollution to the class and discuss the causes of different kinds of air pollution.

2. Lead the class outside, and pass out pencils and white construction paper. Have each group walk around the outside of the school and select a location (such as on ledges, shelves, near air vents, or close to a road) for testing the air. The location should be at least 1.5 m above the ground. Write the names of group members and the chosen location on the paper. (Older students can draw a grid on the paper to facilitate actually counting particles in step 5.)

3. Cover one side of the paper with a thin coating of petroleum jelly. Cut out a circle from the sheet of acetate about 5 cm in diameter, and press this circle onto the coated paper (the petroleum jelly should hold it in place).

4. Place the paper at the chosen location in a vertical posi-

Materials and Equipment:

The whole class will need:

Pictures of air pollution and its effects

A large jar of petroleum jelly

Each group of students will need:

White construction paper approximately 15 cm x 15 cm

A small piece of acetate

Scissors

Tape

String

A pencil

A hand lens

Q Can you see pollutants in the air around you? If not, does this mean there is no pollution in the air?

Water, Stones, & Fossil Bones

Q Why are some locations more polluted than others? Are they near a road, a smokestack, or a vent? Can you think of other types of solid material that might be observed and counted in the same way (such as pollen counts)?

The Author

M. Elizabeth Partridge is an associate professor at Southeastern Louisiana University in Hammond.

tion by taping it or suspending it with string. This will minimize the amount of dirt that might fall onto the paper which is not really air pollution. You might place a sign stating, "Do Not Disturb—Scientific Experiment!" near the paper.

5. After 24 hours retrieve the paper, remove the acetate circle, and inspect the paper with a hand lens. You should be able to see evidence of the residues from air pollution by comparing the area covered by the acetate with the uncovered area. Have the students compare their papers, noting which locations had the most air pollution. (Older students can count the number of particles in two or three squares of the grid drawn on their paper and calculate an average number of particles per square, then compare the averages from different locations.)

Further Challenges:

You might invite an allergist to discuss the effects of pollution on different individuals. Have your students write to the U.S. Environmental Protection Agency's Public Information Center for information on how to prevent air pollution.

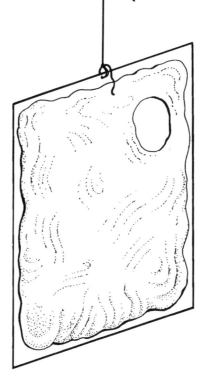

Acid Precipitation

BY MICHAEL J. DEMCHIK

Focus:
The atmosphere contains various particles and gases. The solid particles serve as condensation nuclei on which droplets of water can condense to form rain. Dissolved in these water droplets can be gases. Normal rainfall has a pH value between 5.0 and 5.6, but dissolved gases such as sulfur dioxide and nitrogen oxides, in sufficient amounts, can lower the pH level of the water droplets and increase the acidity of the rain.

Challenge:
What types of particles are found in rainwater or snow? How acidic is the precipitation in your area? Does the pH of the precipitation vary around your county?

Time: 50 minutes

Procedure:
1. Students will conduct the first part of this activity at home. On the morning of a rainy or snowy day before leaving for school, each student should place a Ziploc bag inside his or her jar and fold the top of the bag over the rim of the jar to hold the bag open. (You may want to demonstrate this procedure before the students do it at home.) Place the bag and jar in an open space to collect rainwater or snow, making sure to set the container away from buildings or tall trees which would interfere with falling rain or snow.

The students will leave their precipitation collectors outside until the next morning. If it rains a lot, they may want to put new bags in their jars after they come home from school. It is important that they all collect for the same amount of time so that the particle amounts filtered from the precipitation samples can be more accurately compared.

2. The following morning before leaving for school, the students should seal their bags, carefully fit them into their jars, and put the lids on for the trip to school.

3. For each student's sample, fold a piece of filter paper to fit the mouth of the funnel. Then, have students take turns filtering their precipitation samples, catching the filtered water in their jars (see the figure). If the sample is snow, allow it to melt before filtering it.

4. Set aside the filter papers to dry, and then measure the pH of the filtered samples with the pH paper.

Materials and Equipment:
The whole class will need:
Filter paper (one piece per student)
A funnel
pH paper
Hand lenses or microscopes
A map of your county to write on
A rainy day
Each student will need:
A small jar with a lid
A Ziploc bag

Q Do the pH levels of the class's samples vary? Whose sample has the lowest pH? The highest?

Water, Stones, & Fossil Bones

Q What might explain any variations in pH measured around the county? Are there factories near by? Does the number of homes in a given area vary around the county?

Q Which sample appears to have the most particles? The least?

The Author

Michael J. Demchik, Ph.D., teaches chemistry and physics at Jefferson High School in Shenandoah Junction, WV.

5. On a map of your county, write the pH levels of the students' samples at the locations of their homes. Compare the pH levels from each student's sample. Ask if they see any pattern in the readings.

6. Once the filter papers have dried, examine them for particles with a hand lens or microscope. Compare the amounts of particles from all the samples.

Further Challenges:

At the school, collect samples of precipitation for a month, at approximately the same time of day, and make a plot of pH versus Day of Collection. Does the pH vary over a month's time?

Language Development and Reading

An often-heard question in many elementary schools is, "Why take the time to teach science to elementary students?" Many teachers believe they cannot afford to take the time to teach science. The answer is, "You cannot afford NOT to teach science." Research results leave no doubt about the importance of learning through activity, disclosing that hands-on science programs contribute to the development of language and reading skills.

Students' language and communication skills—talking, writing, performing, drawing, designing, and building—naturally develop from the "action plus talk" of science. In "The Great Flood," for example, many of these skills are used as teammates collaborate in constructing a topographic map and building a terrain.

During "Ancient Earth: A Stratigraphy Cup," the perceptual skills students use to compare the simulated rock layers to road cuts and rocky outcroppings reinforce those needed for reading and writing. Making fine discriminations between objects helps prepare young students to discriminate between letters and words.

To continue development of reading skills and reading comprehension, students need an ability to solve problems and reach conclusions. Science activities, such as "Model of a Seismograph," provide opportunities to determine cause-and-effect relationships and influence the outcome of events. When students can predict the outcome of an action, they gain a sense of control that improves their problem solving skills.

40 Ancient Earth:
A Stratigraphy Cup

BY AMY LOWEN AND LISA WEAVER

Focus:
The order in which the Earth's layers of sediment were laid down is an important clue to the Earth's past. In most sedimentary rock, a given layer will be older (laid down first) than the layer above it. Using this principle and by knowing the type of fossils in the sedimentary rock layers, geologists can determine the relative ages of layers. Rock layers containing the same type of fossils are presumed to be the same age.

Challenge:
Students will make a stratigraphy cup by layering specimens in sand and plaster. Which layers are the oldest? How can you tell?

Time: 45 to 50 minutes

Procedure:
1. Exhibit any fossils (or pictures of fossils) you might have and discuss the different types of fossils. Encourage students to speculate what fossils are and where they were found.

2. Go outdoors, and collect two small specimens to bury within the layers of sand and plaster to represent fossils. The specimens will need to be small enough to fit into the plastic cup. Items such as leaves, sticks, and dead bugs or snails work well. If you do not want to collect specimens, you can use small chicken bones and sea shells. Shell or spiral macaroni, dried beans, or corn can also be used.

3. Have each group put their initials on the bottom of a cup using the masking tape and marker. Put a 1-cm layer of sand in the bottom of the cup and place some of the specimens on top of the sand. Encourage each group of students to put the same types of specimens in the same layer of sand. In this way, a chronology based on the fossil evidence in the stratigraphy cups can be developed.

4. To make the plaster, take two scoops of plaster and put it in the other (empty) plastic cup, add one scoop of water, and stir. Be sure to mix all the way to the bottom of the cup. Pour half of the plaster (pourable, but not too wet) into the plastic cup over the first layer of fossils. Caution students that the layers will disappear if the sand and plaster are mixed together.

5. Carefully add another 1-cm layer of sand.

Materials and Equipment:
The whole class will need:

10–15 pounds of plaster of paris (patching plaster or slaked lime may be substituted)

A plastic scoop such as comes in ground coffee

A 2-L bottle filled with tap water

A 4-L container of sand

Assorted fossils

Each group of students will need:

Specimens to bury such as leaves, sticks, or dead bugs

A wooden stirrer (Popsicle stick)

Two 500-ml, clear plastic cups

Masking tape

A permanent marker

6. Add a second layer of specimens and pour the remaining plaster over them. Set the cup aside to harden. It is best if it hardens overnight and the activity is continued the next day.

7. Examine the plaster and sand layers, and describe any differences among them.

Q Which layers were formed first?

Compare the stratigraphy cup with pictures of layered sedimentary rocks. If possible, view formations firsthand in nearby road cuts or rock outcrops.

8. To examine the specimens in the cup more closely, carefully remove the student-made fossils from the cup by laying the stratigraphy cup on its side and cracking it open by pushing down carefully on the open end. Determine if the specimens have changed; for example, beans, seeds, and corn tend to start to sprout overnight, leaves picked from live trees will dry and shrivel, and some insects may fall apart, etc.

The specimens themselves may have changed but have not "fossilized." The fossil formed is a mold or imprint-type fossil. If the same type of specimens appear in the same layer, for example, all the twigs in the first layer and all the bones in the second, ask, "When did the chicken bones disappear?" (during the earliest period) or "Were twigs common in the most recent plaster period?"

Further Challenges:

To add more fossil layers, use a 1-L plastic bottle with the top cut off instead of the plastic cup. Then fossil types can be layered to illustrate how their periods of existence might overlap. Chicken bones could be found in the last period a certain twig existed, and then continue to be found in more recent periods.

For a take-home activity, have students examine a pile of laundry in their home. From looking at the layers of laundry, students should be able to reconstruct what they wore each day. Ask, "What was the last thing that you wore?" "What did you wear yesterday?" Explain that geologists do this with layers of rocks.

The Authors

Amy Lowen is Director of Education and Collections at the Museum of History and Science in Louisville. Lisa Weaver is Special Programs Manager at the Museum of History and Science.

References:

Crocker, B., and Shaw, E. L., Jr. (1985). Blueberry trilobites and peach strata. *Science and Children*, 22(8), 10–11.

41 The Half-life and Times of Geologic Materials

BY KEVIN D. FINSON

Focus:

Dating of geologic materials sometimes requires the use of radioactive elements that have known rates of changing (decay) from radioactive to non-radioactive material. This rate of change for the substance is called its half-life.

If scientists know how many radioactive atoms are now present in a structure and can measure the number of decayed radioactive atoms, they can calculate how many radioactive atoms were originally present and thus the time that has passed since the material was originally formed.

Challenge:

What does half-life mean? How can radioactive decay be used to date geologic materials? From a graph of a radioactive element's decay rate, can you predict the age of geologic materials containing that element?

Time: 40 to 50 minutes

Procedure:

1. Begin by explaining to students that rocks and other geological materials receive all of their radioactive elements when they are formed. After formation, the radioactive elements begin decaying, giving off radiation and slowly turning into non-radioactive, more stable forms. Scientists know that the rate of decay is different for each type of radioactive atom. The known rate of decay for each element is called its half-life. Half-life is the length of time it takes for half the specific radioactive element to become non-radioactive.

2. Give each group a box with 50 pennies. Place all the coins so they lay heads up in the box. These pennies represent radioactive atoms.

3. Place the lid on the box and shake the box for five seconds in an up and down motion only. Then open the box and remove all the coins that are heads down. The pennies that have flipped over (heads down) represent decayed atoms.

4. Count the number of coins remaining in the box and record this number on a chart as Time Period 1. (Time Period 0 was the time before they shook the box at all.)

5. Replace the lid and shake the box again. Continue repeating the procedure until no pennies are left heads up, each time recording the number of pennies remaining heads up.

Materials and Equipment:

Each group of two or three students will need:

50 pennies

A shoe box with lid

Graph paper

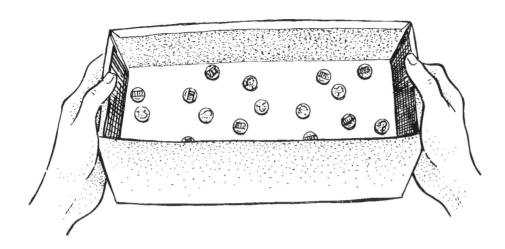

6. Now, repeat the entire procedure, beginning again with 50 pennies heads up. For each time period, add the number of pennies remaining heads up from the first trial with those from the second trial (Trial 1, Time Period 1 + Trial 2, Time Period 1 = Total) so that students are dealing with a total of 100 pennies.

7. Next, prepare a graph with the vertical scale representing Percent Radioactivity Remaining and the horizontal scale representing Time Periods. Each penny equals one percent since a total of 100 pennies was tabulated. Have each group of students plot their own points and then graph a line that would be the result if the "decay" was perfect: Time 1, 50; Time 2, 25; and so forth. Students could plot a line for the class data as well. This would require students adding everyone's trials (all Trial 1, Period 1 data, for example) and determining an average for each time period.

Q What is a half-life? What percent of the original radioactive material would remain after one half-life? Two half-lives? Four or five?

Further Challenges:

Suppose each half-life was 1,600 years. How many years must pass before the amount of radioactive material is reduced to 25 percent? If the half-life is 40 minutes, how much radioactive material will remain after 200 minutes?

References:

Alexander, G. M. (Ed). (1983). *Earth science.* Glenview, IL: Scott, Foresman, and Co.

Ramsey, William L., Phillips, Clifford R., & Watenpaugh, Frank M. (1979). *Modern Earth science.* New York: Holt, Rinehart, and Winston.

American Geological Institute. (1984). *Investigating the Earth* (4th ed.) (Teacher Edition). Boston: Houghton-Mifflin.

Janes, J. R. (1974). *Earth science: Searching for structure.* Toronto: Holt, Rinehart, and Winston.

The Author

Kevin D. Finson is an associate professor of science education at Western Illinois University in Macomb.

42 Timelining
How Old Is Old?

BY CHARLES R. AULT, JR.

Materials and Equipment:

Each group of two or three students will need:

Approximately 15 objects, including photographs, natural history items, historical artifacts, and geological specimens. For example:

- A baby picture or picture of self
- A fresh flower
- News clippings
- An antler or skull
- A cross-section of wood from a tree
- A seashell
- An antique photo
- Pre-1945 rusty tools or household items
- An old book
- A picture of an ancient pyramid
- A picture of a dinosaur
- A layered rock (sedimentary)
- A volcanic rock (igneous)
- Clay
- Sand
- A coarse-crystalline rock
- A fossil plant
- A fossil invertebrate

Butcher paper or other roll of paper

Focus:

An understanding of geologic time has two aspects: duration and chronology. Students' understanding of duration builds on what *short time* and *long time* may mean or their notions of *old*. Their grasp of chronology is revealed by how they apply the concepts of *before* and *after* in arranging objects by age in a single line. This arrangement constitutes a time line.

The concept of age in the context of geologic time may refer to either its durational or chronological aspects. By grouping or spacing objects on a time line to reflect notions of short time, long time, and very long time, students both develop and reveal their grasp of duration. A sense of duration appropriate to historical and geological time is, of course, difficult for students to acquire.

Their grasp of chronology, on the other hand, is developed and revealed by how they arrange objects in a line to reflect a sequence of events. The position represents the time when the object was made and is selected based on the students' notions of *before* and *after*.

If these concepts of duration and chronology are combined, spacing on the line may represent duration. Some students may even suggest that each previous position stands for an event two, several, or even 10 times as long ago as the one after it. Keep in mind that geologic time is so vast compared with the average human life span, or even with recorded history, that a time line in which equal space stands for equal time would be unmanageable.

In this activity students arrange a set of objects from oldest to youngest. This encourages them to think about past periods of time in terms of both duration and chronology (sequence), and gets them to articulate clearly the many degrees of meaning in the word *old*. In doing so, they lay a foundation for acquiring a meaningful conception of geologic time.

Use larger groups or the whole class to discuss the completed time lines. The best insights will emerge from comparisons of different arrangements. This activity has no single, correct solution. Do not be concerned with misconceptions or misinformation about the scale of geologic time in an absolute sense of millions, tens of millions, or hundreds of millions of years. Focus on the more important goals: the development of serial ordering skills, construction of categories of past time, and knowledge of how objects may record information about events in time.

Challenge:

How would you arrange objects and pictures representing your life in relation to geological and historical events? What clues do you use to decide how old an object is? How will you represent different lengths of time?

Time: 30 to 45 minutes

Procedure:

1. Ideally, each student would have his or her own three-meter-long roll of butcher paper to arrange the objects on, but groups of two or three can work together.

2. Introduce all objects in the collections by asking students if they know what the objects are. If an object proves very puzzling and is clearly unknown, discard it. Accept reasonable disagreement over the identity of an object (multiple interpretations are fine); for example, students might not recognize a woodworker's planing tool or the imprints of a frond of a fossil fern. Instruct students to be careful in handling antiques and any precious photographs.

3. Ask the students which object in their set they believe to be the oldest. Discuss their choices (accepting several possibilities) as a way to clarify what *oldest* may mean. You may need to tell the students that the age of an object should be thought of as the time since it was made. Otherwise, they may think a fossil is quite recent because "someone may have found it just yesterday."

4. Ask which object might be considered the youngest. Now, have the students place these objects at opposite ends of the paper.

5. Have the students put their baby pictures at a place on the paper they wish to represent the time of their birth. Point out that the length of the line from that picture to the end

Water, Stones, & Fossil Bones

labeled "Present" stands for their age. Write students' birthdays on the paper, and then ask them how they would space objects on the rest of the line to indicate different amounts of time.

6. Most students will need some guidance at this stage. A first step is to make two groups of objects: young and old. The students can, in turn, split each new group in two and continue refining the arrangement. If deciding on a single series in time for their objects is too frustrating for them, accept partial solutions; for example, arranging the objects in categories such as recent, old, and very old.

Primary grade students may consistently place the baby picture at the beginning of the line. Some will make all time segments equal in length with respect to their birth. A few intermediate and upper elementary grade or middle school students will quickly grasp the problem of representing hundreds or millions of years on the scale used for the 10 years or so standing for their age. As a solution, have them make equally spaced marks going back in time stand for increasingly larger amounts of time. Some students may indicate that each interval may stand for twice the amount of time as the previous one. Others may suggest tenfold intervals—six steps to a million years ago.

Confusion inevitably arises over what the words *age* and *old* may mean to individual students. For example, one student may imagine that a fossil must be very old. Another may decide that the fossil animal probably lived for only a few years and therefore should be categorized as young. Students in the early grades will think about the age of geologic specimens in many, seemingly inconsistent, ways: a rock may be young because it broke off from a boulder yesterday or volcanic rock from a recent flow may be old because it came from an old volcano. Working through these confusions to shared understanding will be of great value.

7. Depending on the ages of the students, you may conclude the activity by discussing the following:

- Reasons for considering objects as young or old
- Divisions between objects from a short time ago, a long time ago, and a very long time ago
- Categories of past time based on how objects were grouped prior to arranging them in sequence
- Criteria for deciding relative age
- Estimates of amounts of past time

• Schemes for dividing the time line to represent duration

Further Challenges:

Continue to add objects to the time line collection. Invent original names for periods of time; for example, the fossil rocks period, the rusty antiques period, the recent life period, and so on. Refine the methods for making length stand for duration, perhaps using different colors of paper or marker for different lengths of time.

This activity might lead students to investigate how geologists categorize time and determine the relative ages of rocks. Search for examples of metaphors of time's vastness, for example, Mark Twain, as quoted by Stephen J. Gould (1987), once wrote, "If the Eiffel Tower were now representing the world's age, the skin of paint on the pinnacle-knob at its summit would represent man's share of that age, and anybody would perceive that that skin was what the tower was built for."

References:

Ault, Charles R., Jr. (1982). Time in geologic explanations as perceived by elementary school children. *Journal of Geological Education, 30,* 304–309.

Bullock, Merry and Gelman, Rachel. (1977). Numerical reasoning in young children: The ordering principle. *Child Development, 48,* 427–434.

Eicher, Donald L. (1968). *Geologic time.* Englewood Cliffs, NJ: Prentice Hall.

Gould, Stephen J. (1987). *Time's arrow, time's cycle: Myth and metaphor in the discovery of geological time.* Cambridge, MA: Harvard University Press.

Hume, James D. (1978). An understanding of geological time. *Journal of Geological Education, 26,* 141–143.

McNeil, Mary Jean. (1975). *How things began.* London: Usborne Publishing, Ltd.

Toulmin, Stephen and Goodfield, June. (1965). *The discovery of time.* London: Hutchinson & Co.

The Author

Charles R. Ault, Jr., Ph.D., is an associate professor of education at Lewis and Clark College in Portland, OR.

43 Magnetic Tracks

BY BETTY B. GRAHAM

Focus:
The Earth's magnetic field has proven an invaluable source of clues for geologists trying to reconstruct the history of the Earth's crust. Certain rocks, volcanic lava among them, contain an iron mineral called magnetite. As they cool from the molten state, these rocks become weakly magnetized in the Earth's magnetic field. Each iron atom acts like a tiny compass needle and will align itself with the Earth's magnetic field.

Materials and Equipment:
The whole class will need:
A warming tray
A directional compass
A sample of magnetite

Each group of students will need:
A plastic coffee can lid
Paraffin
Iron filings
A bar magnet
Safety goggles

✚Safety Note:
Be sure students wear safety goggles when they are heating the wax.

Challenge:
Can you make a picture of the magnetic field produced by a bar magnet? How can magnetized rocks help us reconstruct the Earth's history?

Time: 90 minutes

Procedure:
1. Place chips of paraffin in the plastic lid, and put it on a warming tray until the wax melts.

100   National Science Teachers Association

2. Once the wax is melted, sprinkle iron filings into it, and carefully remove the lid from the warming tray. Immediately pass the iron filings and wax over the magnet so the filings align with the magnet's field, and set them aside to cool. The solidified wax holds the iron filings in place, simulating magnetic particles in molten rock aligning with the Earth's magnetic field as it cools.

Alternate Procedure:

1. As an alternate procedure, pour iron filings onto the shiny side of a piece of wax paper which is on top of a piece of heavy cardboard.

2. Pass the magnet under the filings to make a magnetic field picture.

3. Carefully lay another piece of wax paper, shiny side down, on the picture and iron the two pieces of paper together.

Further Challenges:

How can magnetic particles be evidence to support the plate tectonic theory? How might the lids be rearranged to represent tectonic plates?

References:

DeVito, A., and Krockover, G. (1991). *Creative sciencing: Activities for teachers and children.* Glenview, IL: Scott, Foresman and Company.

Raymo, Chet. (1983). *The crust of our Earth: An arm-chair traveler's guide to the new geology.* Englewood Cliffs, NJ: Prentice-Hall.

Weaver, Kenneth F. (1967). Magnetic clues help date the past. *National Geographic, 131*(5), 696–701.

Q How does the temperature of the wax compare to that of molten rock? What effect would impurities in the rock have on the results? Impurities in the wax? What would you predict about the magnetic fields of rocks from adjoining tectonic plates? How can you find whether or not your predictions are correct?

The Author

Betty B. Graham teaches science at E. J. Martinez Elementary School in Santa Fe, NM.

44 What Is a Fossil?

BY MILDRED MOSEMAN

Focus:

As groundwater moves through sedimentary rock, it can dissolve the bone or shell remains of animals encased in the rock. All that is left is an impression the same shape as the object that was dissolved. This type of fossil is called an imprint. The tracks, tunnels, and trails left by prehistoric animals can also be preserved as imprints.

A second type of fossil, a cast, is formed when the imprint is subsequently filled with minerals deposited from water forming a replica of the bone or shell.

Challenge:

Can you make your own fossil cast with just a seashell, some clay, and a little plaster of paris?

Time: Two class periods

Procedure:

1. Spread a thin coat of petroleum jelly on the seashell and press it into a piece of clay. (This makes for a much cleaner mold.) Remove the seashell and place the clay impression into a paper cup (see the figure below).

Materials and Equipment:

Each student will need:

One or two small seashells
Modeling clay
2 paper cups
Plaster of paris
Water
Petroleum jelly
A plastic fork

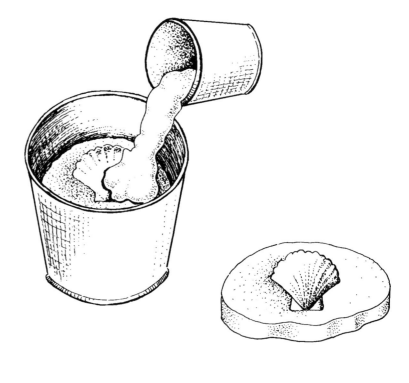

2. Half fill a second cup with plaster of paris, add water gradually and stir until the mixture becomes thick and creamy.

3. Pour the plaster over the clay, and the next day, separate the clay from the plaster.

References:
Winkler, A., Schacter, M., Bernstein, L., and Wolfe, S. (1974). *Concepts and challenges in science.* Book I. Fairfield, NJ: CEBCO Standard Publishing.

Q Are your homemade fossils molds or casts?

The Author
Mildred Moseman, currently retired, taught at the Lincoln School in Sioux City, IA.

45 Future Fossils

BY GERALD WM. FOSTER

Focus:

Webster's New World Dictionary defines a fossil as "any hardened remains or traces of plant or animal life of some previous geological age preserved in the earth's crust."

Scientists generally agree that fossils can be created by four different processes. One process involves animal remains such as shells, bones, and teeth that leave impressions in beds of sand or sticky mud which then slowly turn into rock. In another process mud fills an object, such as a shell, and then hardens into stone. The object then dissolves or decomposes, leaving a stone cast.

The other two processes involve the formation of fossils from softer animal or plant parts. The plant or animal is covered with mud which slowly hardens. Mineral-laden water traveling through the hardened mud seeps into the plant or animal material. Bit by bit minerals fill the spaces in and around the dead cells of this material and harden. This process is referred to as petrification. Finally, other fossils are created by a very slow decay of skin, flesh, leaves, etc. that become thin black sheets of carbon. Scientists can use these fossil remains to understand the type of prehistoric animals and plants that once existed, whether they lived on land or in water and the type of climate that supported them.

Challenge:

Can you tell what kind of object made an imprint in a material? Describe the characteristics of the medium that would make the best fossil imprint. How do the characteristics of the object affect the type of imprint it makes?

Time: 40 to 60 minutes

Procedure:

1. Have students bring to class any examples of real fossils they can find or may have already collected and pass them around the class.

2. Spread out newspapers to cover the work surface. Take several small objects (the objects should vary in pliability and hardness) and make imprints in a small amount of clay. Use a different piece of clay for each object. The greater variety of objects your students use, the more likely they will be to understand why only certain kinds of plants and animals left behind abundant fossils.

Materials and Equipment

The whole class will need:

Real fossils

Each student will need:

Objects that can be used to make impressions and imprints such as:

- Paper clips
- Washers
- Key chains
- Twigs
- Leaves
- Sea shells
- Keys
- Coins

Different kinds of clay such as:

- Play-Doh
- Plasticene
- Floral or potter's clay
- Flour and salt mixture
- Plaster of paris
- Sand
- Dirt
- Clay

Water

A shallow container

Newspapers

Q Do you know how fossils are formed?

3. Divide the class into groups of four or five students each, and then have the students pass their clay impressions on to the other members of their group to guess what object made the imprint.

4. Repeat steps 2 and 3 using other types of clay.

5. Try other media which can make more permanent, fossil-like imprints such as plaster of paris or a flour, salt, and water mixture, or try using different kinds of soils such as sand or clay with water. As you mix water with these various media, maintain a soft, clay-like mixture.

6. To wrap up this activity, take a walking field trip to search for evidence of natural or human-made fossils (such as footprints). Keep written and photographic records of the fossils found in different places. Students can make up stories about the conditions under which the fossils were made.

Further Challenges:

Try to create mixtures that will give clear imprints of objects when they are buried in the mixture. Choose one or several media, for example, sand and plaster of paris could be mixed together with water or sand and glue. Bury several objects with the mixture in a paper cup, and let the mixture harden. Peel away the paper cup from the mixture, and use a nail or some other object to excavate the objects and their imprints. Note the detail and depth of the imprint left by the objects. Students can compare their mixtures and the imprints left by the objects. What mixture made the best fossil medium?

Other discussion questions might include the following: What kind of fossilized remains of our society will future generations find? What evidence do we now have of past generations? Where can you find evidence left by past generations? Can you find natural fossils in human-made structures?

References:

Krockover, G.H. (1986, Winter). Excavating. *CESI News*.

Krockover, G.H. (1986, Winter). Fossils are Fun. *CESI News*.

Q Which of the following objects are likely to form a clear imprint in the clay: peanut shell, grape, penny, or a banana peel?

Q What kinds of objects leave the clearest imprints? Which media best hold an imprint? How clear is a paper clip or twig imprint in modeling clay compared to one left in a sand/water medium?

The Author

Gerald Wm. Foster, Ph.D., is an associate professor of Science Education at DePaul University in Chicago.

Water, Stones, & Fossil Bones

46 Fossil Beds

BY DAVID R. STRONCK

Focus:

Fossils are impressions, traces, or remains of dead plants or animals that are preserved in rock. After scientists find fossils, they work on the problems of interpreting their origin. They try to answer such questions as: What did the living animal or plant look like? What did it eat? How did it move? Why did it die?

Challenge:

Can you reconstruct the skeleton of an animal from its "fossil" bones and describe what it was?

Time: 45 minutes

Procedure:

1. This activity should follow one on human bones. For example, the students could learn how to name major bones shown on a cardboard Halloween skeleton and to "feel" these bones in their own bodies.

2. Obtain the bones of enough small animals to provide one skeleton for each box. Perhaps the easiest bones to collect are those from whole fryer or roaster chickens. Other possibilities are turkey, if your boxes are large enough, and fish. The key is to have all or most of the skeleton. Simply collect the bones after eating and remove any fat or skin by boiling the bones in soapy water.

Materials and Equipment:

Each group of two students will need:

A large cardboard box with a bottom surface of at least 30 cm x 30 cm

Sand (or loose soil) to cover the bottom of the box with a thickness of 2.5 cm–7.5 cm

Bones of a chicken or other small animal

A sturdy plastic fork

A magnifying lens (optional)

A picture of the animal whose skeleton is being reconstructed

✢Safety Note:

Provide plastic forks for the students to dig up the bones as there is a slight risk of them injuring themselves on sharp bones if they use their hands to do the sifting.

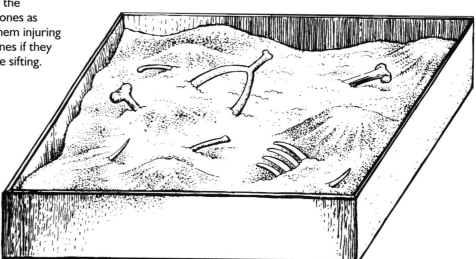

3. Scatter the bones of one animal over the surface of the sand or soil in the bottom of a box, and then carefully bury them. Middle school students may be able to manage the

activity by having the bones of two different animals in the same box.

4. Discuss how scientists know so much about dinosaurs, explaining that all of our knowledge about dinosaurs has come from scientists digging bones out of rocks. Scientists then organize the bones by following the model of animals that are now living.

5. Carefully dig up the bones from the box with a plastic fork. (If they use their hands they might cut or puncture themselves on any sharp bones.)

6. Working from a picture of the animal species whose bones are buried, arrange the bones in the form of that animal. Tell the students that there is one "fossilized" animal in each box. As an option, they may examine the animal's bones, especially the small ones, with a magnifying lens. Under magnification, the students may discover such things as the long bones are hollow and the ends of the bones have cartilage. After finding all the bones, ask the students to describe their animal.

Q How many legs did your animal have? Could it fly? How did it eat? Could it run quickly? How tall was it? How much did it weigh? Why did it die?

If the students want to "preserve" their skeletons, they may glue the arranged bones to a cardboard sheet or a wooden board.

Further Challenges:

Encourage the students to bring to class models of dinosaurs and books about dinosaurs. Some children have stuffed animals that are dinosaurs. Help the children to discuss theories about how dinosaurs became extinct. Perhaps they died of cold and starvation after the Earth's atmosphere was filled with dust or volcanic ash either from a giant meteorite or from volcanic activity.

To explore real fossils, find out if there is an exposed fossil bed near your school and either take the children to dig some fossils or bring some of these fossils to the school to examine and discuss. To find out whether there is a fossil bed near your area, contact the geology department of your local community college or university.

References:

Stronck, David R. (Ed.). (1983). *Understanding the healthy body: CESI sourcebook III.* Columbus, OH: SMEAC Information Reference Center.

Bare Bones Poster. (1984). National Science Teachers Association. Washington, DC.

The Author

David R. Stronck is a professor of teacher education at California State University in Hayward and is currently NSTA Research Division Director.

47 Double-time, Clean Up!!

BY LARRY FLICK

Focus:

We are using the Earth's finite natural resources at an ever increasing rate due, in part, to a continuing increase in the world's population. But that is not the only reason. Our demand for energy to fuel technological progress is increasing at a greater rate than that of the world's population. While population has grown at about 2 percent annually, U.S. demand for energy has grown by about 3.5 percent (Fowler, 1975).

Quantities that increase regularly as a percentage of some base amount are said to increase exponentially. Quantities that increase exponentially will double in size over some fixed interval. To calculate that interval, divide 70 by the annual percent of increase. For example, money earning five percent interest annually will double every 14 years (doubling time = 70/5 = 14). Likewise, oil consumption is growing at about 2 percent a year and will therefore double every 35 years.

Challenge:

Using trash to represent the finite natural resources of coal and oil, how much time and effort does it take to collect at an exponential rate? How often does the amount of "resources" collected double in size?

Time: Three to four weeks

Procedure:

1. Describe to your students the following scenario that revolves around the use of some resource to fuel power plants. The demand for this resource is increasing at an exponential rate, meaning that the demand doubles over some constant period of time.

Here in the United States demand is increasing for electricity. Every year there are new products on the market that require electric power generated by coal, oil, flowing water, nuclear fuel, and natural gas. The result has been a doubling of the amount of electricity used every 10 years. The problem is that every 10 years we must find twice the amount of energy resources (coal, oil, nuclear fuel, natural gas, etc.) used to produce electricity as was used in the previous 10 years (Bureau of Mines, U.S. Dept. of the Interior 1975). Another feature of this 7 percent annual growth rate is that the amount of the natural resources used in a 10-year period is equal to all that was used prior to that time!

2. For this activity, trash from the school grounds will repre-

Materials and Equipment:

The whole class will need:

A designated area in which to collect trash

Containers to be used as standardized volumes of trash, for example cut-off milk cartons or plastic soda bottles

A watch

Each group of students will need:

Large containers for collecting trash such as paper bags or boxes

Gloves

A broom

A dustpan

Paper

A pen

✛ Safety Note:

Be sure all students are wearing gloves when they collect trash, and that all broken glass is swept up, not picked up with the fingers.

sent the resources needed to fuel the power plants. Go out and collect an initial volume of trash to serve as the fuel for the first 10 years. Be sure all students are wearing gloves.

You must establish an initial quantity of trash to be found, the time it takes to find it, and a way of measuring each succeeding (doubled) quantity. Obviously, you should start small. The amounts will soon increase beyond the ability of the students to find them. Starting with 1 L collected on their first round, on their fifth round the students will have to bring back 32 L of trash!

Measuring the amounts by volume is preferred over weight. This will not penalize those who cannot find metal trash. Students will have to consult you in estimating the volume of items too large to fit in the measuring containers.

3. Go out again and collect twice the amount of trash as in the previous round. (Each round represents a 10-year period in the preceding scenario.) It will become increasingly difficult to find the necessary quantity of trash. The amount of time needed to find this trash will also increase. Depending on the area being searched you may need to set a maximum time limit.

If the size of the area to be searched seems large enough, the students can be organized into groups representing countries from around the world and given a specific area within which they are allowed to search. This will produce resource-rich and resource-poor countries.

4. Record the amount of time it takes to find the necessary amount of trash. The amount of time should increase at roughly an exponential rate. You will need a way for calling the group together. Appoint a time keeper and set a maximum time for each search. Record the times in the form of a table.

You may want to designate the school's outdoor trash bins as off limits. These could represent the possible undiscovered resources under the oceans or in Antarctica. An alternative approach would be to include these major deposits within some country's boundaries without calling it to their attention. If they discover it, they will suddenly be in a position to sell their surplus to other countries. This OPEC-like situation can lead to several extensions of the activity.

5. When most of the groups have exhausted their resources, the simulation may end. Graph the amounts of time taken to collect the necessary trash for each round (x-axis is the number of the round, y-axis is the amount of time).

Water, Stones, & Fossil Bones

Further Challenges:

Brainstorm a list of resources which are being consumed at increasing rates. How can growth in consumption be managed? What are some of the consequences of limiting growth?

References:

Bartlett, Albert A. (1976–1978). The exponential function. Parts I, II, III, IV, V, VI, VII, VIII. *The Physics Teacher, 14*(7), 393–401; *14*(8), 485, 518; *15*(1), 37–38, 62; *16*(1), 23–24; *16*(3), 158–159. College Park, MD: American Association of Physics Teachers.

Fowler, John M. (1975). *Energy-environment source book.* Washington, DC: National Science Teachers Association.

Lapp, Ralph E. (1973). *The logarithmic century: Charting future shock.* Englewood Cliffs, NJ: Prentice-Hall.

The Author

Larry Flick is an assistant professor of teacher education at the University of Oregon in Eugene.

Hands-on Mapping

BY JOHN B. BEAVER AND MICHAEL G. JACOBSON

Focus:
Students use the concept of center of gravity to find the geographical center of their state, county, or city. Other indirect measurement and estimation techniques are used to estimate the area and perimeter of a mapped region. The skills they develop and practice include observation, inference, measurement, modeling, working with scale and ratio, estimation, and problem solving.

Challenge:
Can you find the geographical center of your state? How can you estimate the geographical area of a state by weighing a map-model of the state? How can you measure the distance of the perimeter or border of your state?

Time: 45 to 60 minutes

Procedure:

1. Prior to class, carefully glue the road map to the corrugated cardboard and allow the glue to dry.

2. Cut out the outline of the state along its borders. Also cut out the legend and a 5-cm x 5-cm square from the area outside the state's borders, and set them aside for use in Activities 2 and 3.

3. Use the map and the other materials in the following activities. You and your class will probably think of variations and extensions.

Activity 1: Finding the State's Geographical Center

1. Fasten the washer to the string and poke the T-pin or other fastener through the map at any point other than the geographic center. Be sure the pin hole is big enough for the map to rotate freely.

2. Hang the string from the pin to produce a plumb line, and mark lightly with a pencil along the weighted line.

3. Move the pin to another point, allowing the map to rotate freely on the pin. Repeat the process five or six times, making a light pencil mark each time (see the figure). The lines will intersect at the geographical center of the state. Students can check the accuracy of their calculations by attempting to balance the map on one

Materials and Equipment:
The whole class will need:

A road map of your state

Corrugated cardboard at least as large as the map

Craft glue or rubber cement

Scissors

A T-pin or other strong fastener

Cord or string

A heavy washer

Bell wire

A metric ruler

A balance scale with weights

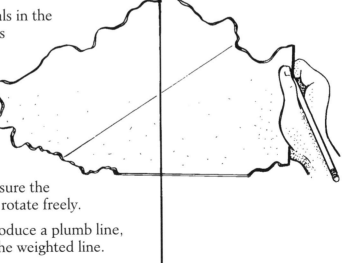

finger at the point they have selected.

Activity 2: Measuring the State's Perimeter

1. Lay a length of bell wire along the edges of the map and carefully shape it to conform to the entire outline.

2. Measure the amount of wire used, and apply the scale of kilometers on the map's legend to calculate the perimeter of the state.

Activity 3: Measuring the State's Area

1. If students have had experience in using a grid to determine the area of rounded objects, they may be able to transfer that experience to the map problem. Weigh the 5-cm x 5-cm square that you cut out of the mounted map.

2. If the 5-cm x 5-cm square weighs 4 g and represents 1,600 km^2, you can weigh the entire mounted map and calculate its area approximately. Students can do this if they have had experience in working with ratios.

Further Challenges:

Have students use the bell wire to estimate the length of river systems, highways, and other linear features on the map. How much of the state's border is land? How much is water? How much is bordered by other states? What is its longest river? What is the lake with the longest shoreline? How far is it from one corner of the state to another? How wide is the state at its widest point? How long?

Students can use the highways and the scale to plan the most efficient route between any two points in the state. Challenge them to plan a highway system connecting major cities. They can use blank outlines of the state for this purpose.

Adapted from Beaver, John B. and Jacobson, Michael G. (1988). Hands-on geography. *Science Scope,* 7(5), 26–27.

References:

Beaver, J. and Jacobson, M. G. (1988). Hands-on geography. *Science Scope,* 7(5), 26–27.

The Authors

John B. Beaver is an associate professor of science education and Director of the Science Education Center at Western Illinois University in Macomb. Michael G. Jacobson is an associate professor in the department of elementary education and reading at Western Illinois University.

The Great Flood

BY CAROLE J. REESINK

Focus:

Using materials easily obtained from a grocery store, students will come to understand the nature of topographical maps and will practice the visualization skills needed to use these maps. The procedure in this activity is not, of course, the actual method used to construct topographic maps, but it simulates how a map could be made if there was a "Great Flood."

A topographic map is a two dimensional representation of a three dimensional landscape. Using a topographic map one can determine where hills and valleys are located and their elevations. A topographic map takes a normal road map a step further in information. Not only can one find various roads, streams, towns, and landmarks, but the topographic map allows us to "see" the elevation of the terrain as well. Many people use topographic maps, from vacationers to city planners. With a topographic map, hikers and cross-country skiers can locate their positions and plan a route of travel because they know, from looking at the map, where the steep hills, cliffs, and canyons are. Geologists and naturalists make extensive use of topographic maps to locate natural features of the terrain. Road planners and city planners also use topographic maps, designing road paths and locations of city facilities.

Challenge:

How can topographical maps help us decide where to build roads and trails? Could topographic maps help us find a good location to build a town or resort? If we were on a hike, could a topographical map help us decide where the easiest path lies?

Time: Several days to several weeks, depending on the depth of study desired

Procedure:

One-hill Terrain

1. Using about one-fourth of the modeling clay, build a single hill on the bottom of the plastic box, making sure that the hill is not taller than the box. For this first activity, instruct the students not to build arches, overhangs, or volcanoes with craters, nor should they put in vegetation or buildings.

2. Place a strip of masking tape vertically on the outside of

Materials and Equipment:

The whole class will need:

A metric ruler

Newspapers

Sponges

A mop

Each group of students will need:

A clear plastic box with a lid (22.5 cm x 17.5 cm x 6 cm) (These may be obtained from the produce or bakery departments of your local grocery store and are very inexpensive or even free if your grocer feels generous.)

An acetate overhead transparency

An overhead transparency marking pen

Onion skin paper for tracing

A metric ruler

Masking tape

A pencil

500 g non-hardening modeling clay

A simple topographic map (figure 2)

Water to fill the plastic box

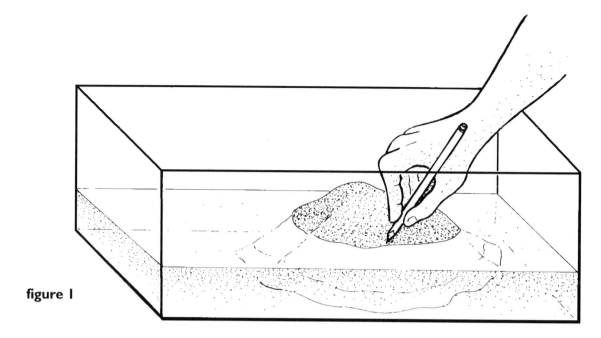

figure 1

the plastic box and mark off 1-cm increments on the tape, starting at the bottom. Next, carefully pour water into the box until the water level reaches the first mark on the piece of masking tape.

3. Using a pencil point or toothpick, make a groove in the clay at the water line, making sure the groove completely encircles the hill (see figure 1).

4. After making the groove along the first shoreline, add more water until the water reaches the second mark on the masking tape and etch a groove along this second shoreline. Continue filling the box with water and marking shorelines until the hill is totally submerged or "flooded."

5. Next, carefully pour the water out of the model while leaving the clay hill in the box. Then, place the lid on the box and tape a piece of clear acetate on top of the lid.

6. Using the overhead transparency pen, carefully trace all of the grooves in the clay while looking directly down on the hill. Try closing one eye and keeping the other eye directly over the pen. Trace the centermost contour lines first and work outward, making sure to trace all of the grooves. Finally, trace the acetate drawing—the new topographic map of the hill—onto a piece of onionskin tracing paper to keep.

7. On the masking tape side of the box, mark a scale of 1 cm = 100 m, starting at the bottom with zero. Now, on the topographic map mark the correct elevations on the contour lines. Mark an "x" on the highest elevation.

Although the students may have been exposed to scales in math and geography, some instruction in this area may be needed. Make the scales as simple as possible, using multiples of 10 or 100 for elevations. Note that with topographic maps, two scales are used—one is the horizontal scale used

Q What is the highest elevation in meters? If it falls between two lines, can you estimate the elevation (write the estimation by the "x")? Do any of their lines cross? Could they ever cross? Is the space between contour lines from a steep hill different from those on a gently sloping hill? How?

on ordinary road maps, and the other is a scale for vertical elevations.

This activity ties into math activities involving coordinate graphing and three-dimensional graphing. If students do not fully understand the first simple-hill map that they make, a second one may be necessary before attempting the three-hill system that follows.

Three-hill Terrain

8. Work in groups of two or three combining everyone's clay to build a three-hill terrain.

9. Repeat steps 3 through 6.

10. Find a place where a stream might flow downhill. Notice the "v" pattern of the contour lines, a characteristic of stream beds.

Building From a Topographic Map

11. From the simple topographic map provided by figure 2, reconstruct the terrain in the plastic box using the modeling clay. One approach is to first roll out the clay into layers 1 cm thick.

12. Cut out the paper map, and lay it on top of the clay. Then cut this contour shape out of the clay. This is the lowest elevation level.

13. From the same cut-out map, cut around the next highest contour line, and repeat step 12, placing this new clay contour on top of the lowest-elevation clay contour.

14. Continue this procedure until all the elevation levels have been cut out of clay and a wedding-cake type of terrain has been built.

An alternative approach is to sculpt the terrain from a mound of clay. It is helpful to place graph paper, to serve as a grid, over the map and another one under the plastic box so the contours can be more accurately

Q What is the elevation in meters of the highest hill, the next highest, and the next? How do you know where the low places are? Are smaller rings at the top or bottom of the hills? Can you tell where the hill is steep or gently sloped?

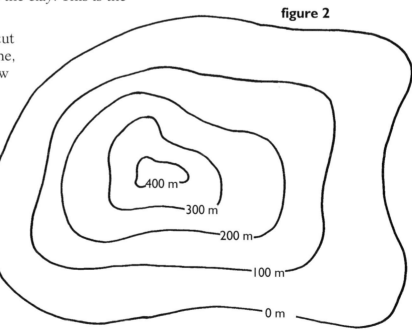

figure 2

Water, Stones, & Fossil Bones

reconstructed. And to accurately measure elevations in the clay, poke a thin wire or pick-up stick into the clay to check how thick the clay has been built up.

Students come up with various methods for replicating the terrain from the map, but these are the two most frequently used.

15. Once the terrains are finished, repeat steps 3–6, and compare the student-made maps to the original map to see how accurately they replicated the terrain.

Q Is the peak in the same location as the original? Does the peak have the same elevation as the original map? How could you make more accurate models?

Further Challenges:

As an extension, build other geologic formations such as volcanoes, canyons, lakes, rivers, cliffs, and overhangs and make additional topographic maps. Find places on the map where a stream could flow down the hillside. Do the contour "v" lines on the map point upstream or downstream? How does a volcano's crater appear on the map? Were there any difficulties in mapping the crater? (Hint: a hole needs to be punched at the base of the crater so that water can flow into the crater, otherwise the crater only fills when the water level reaches the rim of the crater.)

Based on a simplified commercial topographic map (available from science suppliers), create a set of question cards. Some questions to ask include: Can you find the highest point on the map? The lowest? If you walked from Point A to Point B would you be walking uphill, downhill, or on level ground? Is it a steep or gentle hill? Can you find a stream? What direction is it flowing? Where might you build a bridge?

References:

Mapping. (1971). *Elementary science study teacher's guide.* Oklahoma City: McGraw-Hill.

The Author

Carole J. Reesink is a professor of education at Bemidji State University in Bemidji, MN.

How Much Is a Million?

BY LARRY FLICK

Focus:
The study of an object as large as the Earth often involves very large numbers. Encyclopedias and almanacs offer a wealth of information about our planet and the people who inhabit it, using numbers in the millions, billions, and even trillions. What a challenge for the mind to imagine what a million of anything looks, feels, or even tastes like. An important question in teaching Earth science lessons is whether students conceptualize the relationship of these large numbers to anything that is familiar to them.

Challenge:
How can you prove to yourself and others that you have close to one million of something?

Time: 30 to 60 minutes

Procedure:
1. Arrange to have large quantities of several items in your classroom. Use the imaginations of your students to make this step inexpensive and convenient. The items could be found in the school yard or classroom or students might arrange to bring items from home.

Given the availability of good magnifying devices, students may suggest and successfully count grains of sugar, salt, or colored sprinkles. Home may also be a good source of measuring devices that could add to the richness of the activity. For measuring volume, students might bring in measuring utensils such as empty food containers with the volume marked in milliliters.

2. Place all the items to be counted and any measuring devices on tables. Explain that each student will have to find one million of something and be able to prove it to the rest of the class. The class could be divided into small groups, each group counting something different. They may use any of the materials on the table or other things they might think of.

3. Move around the room with a tape recorder and interview the students as they attempt to count, measure, weigh, or estimate one million of something. If a tape recorder is not available, select several interviewers ahead of time who will take notes. In either case, have prepared questions that probe the methods chosen to count one million objects and the students' conceptualization of one million.

Materials and Equipment:
The whole class will need:
Rulers
Graduated containers
Plain containers
Balance scales
Scrap paper
Microscopes or hand lenses (optional)
A tape recorder and blank tape
Calculators
Objects to be counted such as:
 Holes in ceiling tiles
 Blades of grass per 10 cm^2
 Threads per cm^2 in a fabric
 Letters on a page
 4 kg rice
 4 kg popcorn
 8 kg aquarium gravel
 6 kg dry navy beans
 6 kg dry breakfast cereal
 4 kg wrapped hard candy
 9 kg elbow or shell macaroni

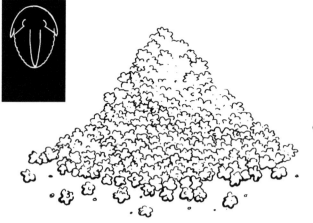

Some sample questions are:

- What was there about the substance you were working with that made you want to use a method other than counting one-by-one?

- Why did you choose this method instead of another?

- How long did it take you to establish that you had a million?

- What was your method and what were some problems you encountered with it?

- What can you tell others about how large a million is now that you have attempted to count a million of something?

- What will you think about the next time you see the word million?

Some possible methods used for determining a million of something might be:

- Counting one-by-one—This method will probably be abandoned quite early. It takes too long and the count is easily lost. However, trying to use this method is instructive.

- Sampling (Volume)—Students take a portion that they have already counted and use it as a standard. For example, to count the number of kernels in a liter jar of popcorn, students first determine how many there are in a small paper cup. They can then find out how many paper cups it takes to fill the liter jar.

- Sampling (Area)—Using a ruler, students make a square inch and count out the number of rice grains that fit into it. Students can then determine how many square inches will represent one million.

- Doubling or Halving—Students count out a certain number of grains or rice. Then they make another pile the same size. They add the two piles together and match another pile to the larger one "by eye." This would be called doubling. Halving would proceed from a larger pile to smaller piles. Students would still count the smaller pile.

- Weighing—Students count out a certain number of peas, for example. Then using a balance, they weigh that amount. They then figure out how much one million peas would weigh.

Students are likely to devise many other techniques. The recording of the process will help document their emerging conception of one million.

4. Use the recorded interviews (or notes) at the next class period to stimulate discussion on what one million looks like. Have the students draw or write down the steps illustrating their thinking processes while they were measuring so that someone else could follow them. These steps might be read aloud in an attempt to "teach" a new method of measuring, and these various approaches could then be compared.

Further Challenges:

How much is one billion? Use a word processor to produce and duplicate a page of 4,000 dots. One billion dots would cover 250,000 pages. A stack of these pages the height of a 41-story building (over 122 m) would contain a dot for every person on Earth (about 5 billion).

How big is the Earth by volume and by weight?

How many liters of oil are consumed each year by each country? What is the current estimate of oil left on Earth?

How many miles of rivers are there in the United States?

How many grains of sand are in a bucket? On a beach?

References:

National Science Teachers Association. (1990). "The Great Lakes in geological time." *NSTA Great Lakes Jason Curriculum.* Washington, D.C.: National Science Teachers Association.

The Author

Larry Flick is an assistant professor of teacher education at the University of Oregon in Eugene.

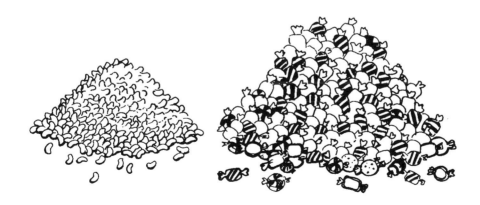

51 Litter Alert

BY ROBERT N. RONAU

Focus:
Litter items gathered from a designated public area can be classified, tabulated, and graphed to discover patterns and trends of littering.

Challenge:
How much litter is discarded in your neighborhood or in your favorite park? Is there a pattern of littering that can be used to estimate levels or trends of littering or to predict future littering at a given site, or at several designated sites?

Time: Five class periods over two to three weeks

Procedure:

1. Select an appropriate site, such as a roadside or public park, that is not regularly cleared of litter. If the site is too large for the class or the group assigned to the site to completely clear it of litter, then carefully, but discretely, mark the boundaries of the area to be investigated.

2. Collect all of the litter in this area. If the time since the site was last cleared of litter is unknown or litter removal from the site is not controlled, this sample would not be appropriate to use as part of the data set. This litter can be used, however, to help determine categories of litter for the site. Some possible sets of classifications might include the following:

Materials: paper, aluminum, glass, Styrofoam, plastic, other

Sources: fast food, household, beer/alcohol, industrial, vehicle parts, other

Effects on environment: years needed to biodegrade, effects on wildlife, health or safety hazard to humans, aesthetic impact, other

Value of collected items: bottles, cans, newspapers, Styrofoam, and non-recyclable items (Note that items are non-recyclable for different reasons; for example, it may still be less expensive to create new products, no reclamation center is nearby, or the by-products of recycling are too toxic.)

Other types of classifications are possible. The same collection of litter can and should be classified into several categories for study.

You may want to introduce information from the Environmental Protection Agency, the American Plastics Institute,

Materials and Equipment:

The whole class will need:

Computer with spreadsheet and graphing programs (optional)

Literature on solid waste disposal

A broom and dustpan for broken glass

Each student will need:

Containers for collecting litter (bags or boxes)

Gloves

Graph paper

✢ Safety Note:

Students should immediately record and classify any broken glass, sweep it up, and put it directly in the nearest trash receptacle rather than trying to collect it for later classification. Be sure all students are wearing gloves when they collect trash. Also, make sure students wash their hands thoroughly after collecting and classifying.

or other sources on the amount of time needed for different items to biodegrade, the dangers to plants and animals of human refuse, and the value of recyclable materials and recycled products.

3. Establish time frames for trash collection at the site including the time between collecting, the days of the week for collection, and the overall duration of the experiment (or number of collections). For example, collections could be made every other day for two weeks or every Monday and Friday for three weeks (this latter schedule isolates the weekend for special study).

4. Collect litter on the specified days and sort it into the categories determined by the class. Be flexible because new types of trash may be discovered that do not readily fit into

Q What are the advantages and disadvantages of different collection schedules?

Q Is there a difference in litter found on different days of the week? Are there categories of refuse prevalent at one site but not another? Does a particular type of litter occur most frequently? Can you formulate a simple solution to diminish this form of litter?

The Author

Robert N. Ronau is an assistant professor of mathematics education at the University of Louisville.

the initial categories. Record the number of items (or weight of the items) found in each category. Tabulate and chart the data as the number of items (or weight) collected versus the category.

5. Repeat the collecting, sorting, tabulating, and charting for the remaining collection days. From these investigations, search for patterns that might uncover trends of littering. A computer spreadsheet is particularly helpful for this step.

Categorize the litter in several ways. Does one presentation create more of an impact than another? For example, does classifying the litter by value have a greater impact than classifying the litter by number of items? What type of graph best conveys the environmental impact of the litter?

Further Challenges:

Predict, from the established pattern, the amount of litter that could be collected from the designated area over longer periods of time (six months, one year, two years, etc.). Assuming that the designated area is typical, estimate the amount of litter that is discarded throughout the neighborhood, park, city, or state. Again, a computer spreadsheet is helpful for this part. To extend participation beyond your class, form a partnership with another class in another school (possibly in another city or state) to compare similarities and differences in littering patterns.

The students could expand their analysis of these littering patterns and discuss the implications for the environment of the predicted littering. The class could devise a plan of action to limit littering in the area, particularly if the area is a sensitive one, and prepare a report to present to state or local officials.

Does your state have a container law? If so, does this affect the types of litter found? Compare your findings with those of a class in another state with laws different from yours.

Water, Stones, & Fossil Bones

Council for Elementary Science International

**Officers
1990–1991**

President
Karen K. Lind
University of Louisville
Louisville, KY

President Elect
Bonnie B. Barr
SUNY Cortland
Cortland, NY

Retiring President
Mark R. Malone
University of Colorado
Colorado Springs, CO

Secretary Treasurer
JoAnne Wolf
Mesa Public Schools
Mesa, AZ

Recording Secretary
Eileen Bengston
Hurlbutt Elementary School
Weston, CT

Membership Chair
Betty Burchett
University of Missouri
Columbia, MO

NSTA Representative
Joan Braunagel McShane
Jefferson Elementary School
Davenport, IA

Director
Betty Holderread
Santa Fe Middle School
Newton, KS

Director
Larry Flick
University of Oregon
Eugene, OR

Director
Delmar Janke
Texas A & M Univeristy
College Station, TX

Director
Carolyn Petty
Bemidji, MN

Director
Gerald Krockover
Purdue University
W. Lafayette, IN

The Council for Elementary Science International (CESI), an affiliate of the National Science Teachers Association, is an organization interested in improving the quality of science education for preschool through elementary students at the local, national, and international levels.

The purposes of CESI, according to the CESI Constitution, are ". . . to stimulate, improve, and coordinate science teaching at preschool and elementary school levels and to engage in any and all activities in furtherance thereof; to promote the improvement of science progress which begins in preschool or first grade and develops in a continuous and integrated fashion through grade 12 and beyond . . ."

CESI provides a variety of resources in an effort to promote quality preschool/elementary science education. The intended audiences are classroom teachers, resource staff, supervisors, administrators, research personnel, informal educators, and methods instructors. These resources include

- Sessions at international, national, state, and local science teacher conferences and meetings. Presenters include nationally recognized experts in the field of elementary science education and classroom practitioners who share hands-on materials, research data, classroom activities, and inservice suggestions.
- A quarterly newsletter offering teaching suggestions and updates on current issues in preschool/elementary science education.
- Sourcebooks written on specific science topics of interest to preschool/elementary practitioners.
- Awards programs that recognize exemplary elementary science teachers, teachers new to our profession, and principals who are supportive of elementary science instruction.
- International projects that provide global perspectives and methods for sharing mutual interest/concerns.
- A directory of members—people identified as having an interest and voice in preschool/elementary science.
- Monographs and occasional papers on issues of specific interest to elementary science educators and those interested in promoting quality elementary science education.
- Positions on issues and spokespeople for preschool/elementary science education.

- A forum of educators to voice their opinions, share their ideas, and develop a professional comradery with those having similar interests and needs.
- A professional link for preservice, inservice, and postservice science educators.

Water, Stones, & Fossil Bones is the sixth sourcebook compiled by the Council for Elementary Science International. It is the second in a series of content-specific resource books. It is the first joint publication of NSTA and CESI. The sourcebooks follow a similar format, and the lessons emphasize teaching science through content- or skill-based activities.

Previous CESI Sourcebooks

CESI Sourcebook I
Outdoor Areas as Learning Laboratories
edited by Alan J. McCormack

CESI Sourcebook II
Expanding Children's Thinking Through Science
edited by Michael R. Cohen and Larry Flick

CESI Sourcebook III
Understanding the Healthy Body
edited by David R. Stronck

CESI Sourcebook IV
Science Experiences for Preschoolers

CESI Sourcebook V
Physical Science Activities for Elementary and Middle School
edited by Mark R. Malone